SINGING THE TURTLES TO SEA

ORGANISMS AND ENVIRONMENTS

Harry W. Greene, Consulting Editor

1. *The View from Bald Hill: Thirty Years in an Arizona Grassland,* by Carl E. Bock and Jane H. Bock

2. *Tupai: A Field Study of Bornean Treeshrews,* by Louise H. Emmons

3. *Singing the Turtles to Sea: The Comcáac (Seri) Art and Science of Reptiles,* by Gary Paul Nabhan

4. *Amphibians and Reptiles of Baja California, Including Its Pacific Islands, and the Islands in the Sea of Cortés,* by L. Lee Grismer

5. *Lizards: Windows to the Evolution of Diversity,* by Eric R. Pianka and Laurie J. Vitt

6. *American Bison: A Natural History,* by Dale F. Lott

The publisher gratefully acknowledges the generous contributions to this book provided by Drs. David A. Lennette and Evelyne T. Lennette, Agnese Haury, the David and Lucille Packard Foundation, the Kelton Foundation, and the General Endowment Fund of the Associates of the University of California Press.

COMCÁAC CONTRIBUTORS

Amalia Astorga de Burgos
Irma Astorga
María Luisa Astorga de Estrella
Olga Astorga
Santiago Astorga Flores
Victoria Astorga de Barnet
Francisco "Chapo" Barnet Astorga
Francisco "Pancho Largo" Barnet
    Astorga
Ignacio "Nacho" Barnet Astorga
Martín Enrique Barnet
Miguel Barnet Contreras
Ramona Barnet Astorga
Raymundo Barnet
José Luis Blanco Blanco
María Guadalupe Blanco Montaño
Adolfo Burgos Félix
Anita Burgos Astorga
Carmen Burgos de Barnet
María Burgos Félix de López
Rafael Comito Robles
Ricardo Francisco Comito Sesma
Samuel Comito Robles
Elvira Cubillas de Torres
María del Carmen Díaz de Comito
Mercedes Díaz de Barnet
Efraín Estrella Blanco
Ricardo Estrella Blanco
Josefina Ibarra Félix de Martínez
Lydia Ibarra Félix de Blanco
María Félix Molina
Manuelito Flores
María Ofelia Flores Ibarra
Genaro Herrera Casanova
Juanita Herrera de Montaño

Gabriel Hoeffer Félix
Raquel Hoeffer Félix
Alfredo López Blanco
Antonio López Blanco
David López Palma
Emilio Anacarsis López Morales
Guadalupe López Blanco
José Luis López Morales
Ramón López Barnet
Ramón López Flores
Jesús Alberto Martínez Ibarra
José Luis Martínez Ibarra
Solorio Martínez Ibarra
Victor Martínez Ibarra
Alberto Mellado Moreno
Alfonso Méndez
Elena Gastelum Méndez de Torres
Estela Méndez Montaño
Moisés Méndez R.
Rodrigo Méndez Serapho
Rosita Méndez de Méndez
Anabertha Molina Romero
Ernesto Molina Villalobo
Francisco Molina Sesme
José Guadalupe Molina Morales
Malca Molina
María Félix Molina de Hoeffer
Raúl Molina
Roberto Carlos Molina
Saúl Molina Romero
Martina Monrroy
Jesús Montaño
Lydia Montaño de Blanco
Arturo Morales Blanco
Cleotilde Morales Colosio

David Morales Astorga
Fernando Morales Ortega
Humberto Morales
Roberto Morales Blanco
Armida Patricia Moreno López
José Juan Moreno Torres
María del Carmen Moreno López
Patricia Moreno de Mellado
José Guadalupe Ortega Molina
Héctor Perales M.
Ramón Perales
Antonio Robles Astorga
Cornelio Robles Barnet
Enrique Robles Barnet
Israel Robles Barnet
Josuë Robles Barnet
Jesús Rojo Montaño
Norma Alicia Rojo
Felipe Romero Blanco
Francisco "Ruben" Romero Morales
Humberto Romero Perales
Magdalena Romero Morales
Manuel Romero
Pedro Romero
Pedro Romero Astorga
Rafaela Romero
Roberto Romero
Alfonso Rosendo Flores
Ana Luisa Torres
Angelita Torres Cubillas
Daniel Armando Torres Cubillas
Dolores Torres Cubillas
Isabel Torres
José Ramón Torres Molina
Manuelita Torres Cubillas

singing the turtles to sea

The Comcáac (Seri) Art and Science of Reptiles

Gary Paul Nabhan

Undertaken in Collaboration with the Seri Tribal Governors and the Traditional Council of Elders

WITH A FOREWORD BY HARRY W. GREENE

University of California Press   Berkeley   Los Angeles   London

University of California Press

Berkeley and Los Angeles, California

University of California Press, Ltd.

London, England

© 2003 by the Regents of the University of California

Library of Congress Cataloging-in-Publication Data

Nabhan, Gary Paul.
   Singing the turtles to sea : the Comcáac (Seri) art and science
of reptiles / Gary Paul Nabhan.
      p. cm.—(Organisms and environments ; 3)
   Includes bibliographical references and index.
   ISBN 0-520-21731-4 (cloth : alk. paper)
   1. Seri Indians—Ethnozoology.   2. Reptiles—Mexico—Sonora (State).
3. Seri Indians—Folklore.   I. Title.   II. Series.
   F1221.S43 N32   2003
   398'.36979'097217—dc21   2002035363

Printed in the United States of America
10 09 08 07 06 05 04 03
10 9 8 7 6 5 4 3 2 1

Time—the lizard in the sunlight. It doesn't move, but its eyes are wide open. *They love to gaze into our faces and hearken to our discourse.*

It's because the very first men were lizards. If you don't believe me, go grab one by the tail and see it come right off.

CHARLES SIMIC, from "Time—A Lizard"

contents

Maps are on pages 45, 46, and 78

Key to Pronunciation of Words in Cmique Iitom xiii

Foreword, by Harry W. Greene xv

Introduction 1

PART ONE

1 Islands of Uniqueness: Endangered Cultural Knowledge
   of Endemic Creatures 14

2 Mapping the Comcáac Sense of Place: Seri Homelands and Reptilian
   Habitats 40

3 The Shape of Reptilian Worlds: Island Biogeography and the Herpetofauna
   of the Sea of Cortés Region 60

4 Naming the Menagerie: How to Sort One Snake from Another 98

5 Reptiles as Resources, Curses, and Cures: How the Comcáac Recognize Beauty,
   Utility, and Danger 118

6 What Eats from the Turtle's Shell, What the Turtle Eats: Comcáac

  Perceptions of Local Ecological Interactions 154

7 The Comcáac as Conservationists: Practicing What They Preach,

  and Benefiting from Alliances 178

8 The Historic Decline and Recent Revival of Traditional Ecological

  Knowledge 200

PART TWO

9 Accounts of Reptiles Known by the Comcáac 226

Appendix. Reptile Specimen Records from the Sonoran Coast

  and Nearby Islands 283

Literature Cited 291

Index 303

Note on translations in text: Cmique Iitom place-names and phrases are translated contextually, not literally. Where at least one term is untranslatable, no translation is provided. William Bright assisted in preparation of this key; Becky Moser and Steve Marlett kindly edited all terms and translations in the text.

Orthographic Symbol Used by Comcáac	International Phonetic Association Symbol	Phoneme	English Approximation
a	a	low back vowel	igu*a*na
c, qu	k	voiceless velar stop	ge*c*ko
cö	kʷ	voiceless labialized velar stop	*Qu*ammen
e	æ/ɛ	low-mid front vowel	leatherb*a*ck
f	Φ	voiceless fricative	*f*in
h	ʔ	glottal stop	as in sound separating the words *an* and *aim* (versus *name*)
i	[i]	high front vowel	g*i*la monster
j	x	voiceless velar fricative	as in Scottish lo*ch*
jö	xʷ	voiceless labialized velar fricative	*wh*ite
l	ɬ	voiceless lateral spirant	as in Welsh *ll*
m	m	labial nasal	*m*ud turtle
n	n	dental nasal	igua*n*a
o	o	mid back rounded vowel	geck*o*
ö	w̥	vowel	*w*hipsnake
p	p	voiceless bilabial stop	*p*iebald chuckwalla
r	ɾ	voiced alveolar flap	zeb*r*a-tailed lizard
s	s	voiceless aveolar spirant	*s*idewinder
t	t	voiceless dental stop	*t*ortoise
x	χ	voiceless uvular fricative	as in *r* in French le*tt*re
xö	χʷ	voiceless labialized uvular fricative	*wh*ite, pronounced as though gargling
y	j	palatal semivowel	*y*ellow-bellied sand-snake
z	ʃ	voiceless post-alveolar spirant	fi*sh*

foreword

HARRY W. GREENE

Singing the Turtles to Sea is the third volume in the University of California Press's new series on organisms and environments. Our main themes are the diversity of plants and animals, the ways in which they interact with each other and with their surroundings, and the broader implications of those relationships for science and society. We seek books that promote unusual, even unexpected, connections among seemingly disparate topics; we want to encourage writing that is special by virtue of the unique perspectives and talents of the author.

Gary Paul Nabhan's *Singing the Turtles to Sea* explores the special significance of reptiles for the Seri, a small but distinctive group of Native Americans in Sonora, Mexico. This masterful book, replete with fascinating Seri observations of everything from small gray tree lizards to massive Leatherback Turtles, provides a case study in ethnoherpetology. It also illuminates some key issues for the preservation of cultural and biological diversity. Why do reptiles, elsewhere so often reviled rather than protected, play such a central and positive role in the lives of the Seri? Can local communities sustainably harvest globally endangered organisms in a framework that encompasses both conservation and traditional beliefs? And more broadly, what can indigenous peoples teach us about managing and appreciating our surroundings? How can we as outsiders, onlookers, or even participant-observers in the life of a particular place steer between two incomplete and misleading visions: that native peoples always live in harmony with nature, and that exclusive reliance on Western science will save us from the ever more devastating impact of humans on Earth?

The Seri landscape encompasses several major habitats and a diverse fauna. Here arid-adapted species of the Sonoran Desert mingle with arboreal reptiles more typical of tropical thorn scrub to the south, and until recently crocodiles inhabited the rivers and estuaries. Several kinds of marine turtles and a seasnake regularly visit the adjacent Sea of Cortés, and islands in those waters harbor endemic giant chuckwalla lizards and rattleless rattlesnakes. Of course crocodilians and venomous serpents can be dangerous to humans, but most lizards and snakes are harmless and easily handled; tortoises, horned lizards, and sidewinder rattlers are the very essence of peculiar—they invite lingering observation, they engender notions of magic and mystery. The Seri have long had close and perceptive relationships with all of these creatures; some kinds they eat, others they keep briefly as pets, and most local species appear in their songs and stories. "Ecological referents," descriptive terms based on the behaviors of particular species, frequently enliven Seri taxonomy, and their beautiful carvings of animals are extraordinarily lifelike in posture. The Seri and their reptile associates are linked in a more disturbing way as well: both are threatened by socioeconomic pressures, technology, and cultural homogenization.

Gary Paul Nabhan brings a special blend of abilities, experience, and inclinations to this project; an unusually broad and widely traveled biologist, he seemingly is interested in everything. Trained as an ethnobiologist, with a Ph.D. from the University of Arizona, Gary now serves as director of the Center for Sustainable Environments at Northern Arizona University. His scientific investigations have emphasized the roles of native foods in the diets of cultures adapted to arid regions, conservation biology, and the importance of nature education in the lives of children. Along the way Gary has edited and written some dozen scholarly and popular books, each of them distinguished by his personal candor, luminous prose, and unabashed delight in life. These accomplishments have gained him a prestigious MacArthur Fellowship, the John Burroughs Medal for outstanding natural history writing, and numerous other honors. Gary has worked with the Seri and other Native American groups for decades, and has been deeply involved in community-based programs to foster environmental education and economic stability.

For those concerned about the fate of biological diversity, Gary Nabhan's final chapter is guardedly optimistic and hence uplifting. The Seri live in a place that is shockingly harsh by some standards and extraordinarily rich by others, and we can ill afford to lose the knowledge of how they do manage to survive. In the face of early-twenty-first-century problems, with an awareness that their traditional lifestyle and the environment are inseparable, the Seri are exploring ways of preserving both. Some of the solutions are relatively simple, they are locally inspired and controlled, and they just might work. Although other books have examined human interactions with reptiles, *Singing the Turtles to Sea* is especially valuable for focusing on a local culture, its involvement with an entire herpetofauna, and their shared prospects for the future.

To recognize the poetic dimension of ethnography does not require one to give up facts and accurate accounting for the supposed freeplay of poetry. Poetry is not limited to romantic or modernist subjectivism; poetry can be historical, precise, objective.

JAMES CLIFFORD, "On Ethnographic Surrealism" (1981)

introduction

José Luis Blanco was ambling down the sandy road, trying to hawk his wares to me and to my friends. Hardly a minute had passed since our truck had come to a halt in Punta Chueca, and yet he was already onto us, intent on being the first Seri artisan to sell a carving to the arriving load of outsiders.

Blanco's stone and ironwood carvings of animals were always innovative, so over the previous three years I had purchased a dozen of his figurines: chuckwallas, iguanas, giant serpents, Sidewinders, and rattlers with toads or horned lizards in their mouths. But on my last visit to this Comcáac village on the Sea of Cortés, I sensed that his neighbors and competitors were jealous that I was buying so many carvings from him. This time I decided to shop around and spread my "business" more widely.

No sooner had I finished whispering this vow to myself than Blanco tugged at my sleeve. He greeted me loudly, his mirror sunglasses flashing at me and his arms waving as if I were his long-lost brother. His wife then shuffled up next to us, carrying a large carving partially hidden beneath a towel.

"Hᴀɴᴛ ᴄᴏᴀ́ᴀxᴏᴊ!" Blanco barked at me, calling me by my nickname in his native tongue, "Horned Lizard." "Hant Coáaxoj, how are you, my fren?" he continued, flatly repeating one of the few English phrases he had memorized.

Whenever we were kidding around and he called me "Horned Lizard," I would in turn refer to him as the "Old Gringo." I would tell his family that the Old Gringo knew more English than I did. He would then motion to his wife to unveil his latest carving. This time it was a lovely, sinuous rattlesnake carved from a dead ironwood branch and polished to a luster with cordovan shoe polish.

"This carving, especial for you my fren, Hant Coáaxoj, especial for you."

I tried in vain to divert their attention, but Blanco's wife placed the new carving in my hands, intent on the idea that I should have it. Blanco then shifted from his few formulaic phrases of English and began to negotiate with me in Spanish.

"Hant Coáaxoj," he whispered emphatically, "I made this one especial for *you*." Seeing that I was not yet buying his line, he grabbed the carving and put it right up in front of my face. "The other snakes you've bought from me, well, they're all eating horned lizards, your namesake. Poor Hant Coáaxoj! Dead again! But this time, well, I save you, my fren."

"What? You save me?" I asked, puzzled.

"I was going to carve Hant Coáaxoj eaten and dead again! But then, the last time you came, you pay me all I ask for. This time, well, I make snake, but see that bulge?"—he pointed halfway down the snake's sinewy length— "This time, Hant Coáaxoj gets swallowed by Snake, but Hant Coáaxoj doesn't die. Hant Coáaxoj lifts his thorny head up and cuts right through Snake's skin to get out!"

"That's why I just see Snake, no Hant Coáaxoj, in the carving?" I asked.

"Just the bulge! I've seen him escape that way, right up through Snake's skin. Hant Coáaxoj, this time you go free!"

"I go free?"

"Yes, you buy this snake from me for a good price, and you go free!"

SOMETIMES it is the Old Gringo Blanco's stories that get to me; sometimes it is Jesús Rojo's songs. But whenever I go to the villages of Punta Chueca and Desemboque, Sonora, or to the fishing camps of Isla Tiburón, I come back with stories and songs reeling through my head. Many of those tunes and tales are about reptiles: Lizards. Turtles. Crocodiles. Snakes. Some are of creatures no longer seen in the flesh as far north as Sonora, but word of them still circulates around the campfires of the Comcáac, as Blanco and Rojo call their own people.

Through the following stories, song fragments, photos, maps, drawings, diagrams, and discussions, I hope to convey the richness of local knowledge about the reptiles remaining within two coastal villages of the Comcáac, or Seri, people.* Theirs, like other indigenous communities' *traditional ecological knowledge*, is a curious mix of scientific insight and artistic expression, "evolving by adaptive processes and handed down through generations by cultural transmission" (Berkes 1999). Whenever I listen to a Seri storyteller narrate about Regal Horned Lizards or Desert Tortoises, I hear little if any break between what we might call their "indigenous science" and their "folk art."

Consider for a moment the possibility that the traditional ecological knowledge about reptiles shared by the Comcáac community has the capacity to enrich our own views of the natural world and, accordingly, to encourage us to better protect natural diversity. If, however, this "folk knowledge" is categorically dismissed as unscientific and ultimately replaced by Western scientific knowledge alone, we stand to lose something of import. Comcáac culture may not be the only entity thereby diminished, particularly if this knowledge can help protect, restore, and recover threatened reptiles. If such indigenous knowledge is not taken into account in conservation management plans, scientists will have failed to draw upon a body of understanding that might help

---

*I will use the term *Comcáac* to refer collectively to the people and their culture, and the term *Seri* to refer to activities peculiar to one or a few individuals within the community, as well as to products advertised to the outside world as "Seri Indian." The Seri Indian language, called Cmique Iitom by its native speakers, is the sole surviving member of its immediate linguistic family. It is tentatively considered part of the putative Hokan superfamily, which includes the Yuman languages of Baja California and adjacent states (Marlett 2000).

them solve problems critical to the survival of endangered species (Nabhan 2000b).

By subtitling this book "The Comcáac (Seri) Art and Science of Reptiles," I join the growing number of folklorists, ethnobiologists, linguists, and cultural ecologists who have slipped from the constraints of disciplinary conventions in their explorations of the natural world. The work I present here, like that of many other ethnobiologists, has been accomplished in collaboration with tribal leaders, teachers, hunters, healers, storytellers, artisans, and para-biologists belonging to contemporary indigenous communities. No one voice, no single scholarly discipline in isolation, can do justice to all facets of these collaborations.

And yet, far too many published ethnobiological treatises fail to capture fully the diverse tones and timbres of contemporary cross-cultural discussions regarding local flora and fauna. Far too often in the past, ethnobiological texts paraphrased, summarized, or homogenized such discussions to the point that the heterogeneity of native speakers' voices could barely be heard. In this book, as I have done in previous ethnobiological essays, I will experiment with a mix of means to capture this heterogeneity. I may in the end declare this experiment a failure, as I have others in the past, but I hope it will encourage colleagues of mine who are similarly inclined to keep experimenting.

My concern to colorfully capture the range of cultural perspectives regarding local wildlife is at once literary, ethical, and scientific. While a growing number of fieldworkers in the natural and cultural sciences now share this concern, only a few ethnobiologists (e.g., Rea 1997; Nelson 1983) have experimented with literary devices to capture this multivalence. Few scientific explorations of northern Mexico's deserts and seas convey much sense of what it really feels like to be in the field there—the exception being John Steinbeck and Edward Ricketts's classic collaboration *The Sea of Cortez*, published in 1941. Most other natural and cultural histories from the central Gulf coast of Sonora have muted the unpredictable seas and the harsh desert weather, the phantasmagorical landforms and equally bizarre life-forms. They hardly howl, whine, or whimper about the humorous, exhilarating, or exasperating cultural encounters that inevitably occur whenever any of us leaves the safety of our home.

Perhaps we need more scientific literature that embraces such "thick description" as part of its task. Perhaps we need to foster a brand of ethnobiology that features the expressive individuals who interpret their culture's traditions in idiosyncratic and innovative ways, rather than continuing to stereotype every soul in a community as if they uniformly adhere to the same cultural norms. Perhaps we must further affirm an ethic that has gradually emerged in ethnobiology, one that assures benefits to communities rather than allowing resources to be unfairly extracted from them. I am grateful to my many colleagues in ethnobiology who not only share these views but have pioneered ways in which to live by them.

We need, in short, to see more human faces (of all colors) in our science. Such a science will treat Indian people not as mere objects of research, but as collaborators in understanding and wisely managing the natural world. As award-winning Mexican ecologist Victor Toledo (1995) has written:

> Just as a new literature [of indigenous peoples] is arising, a new type
> of ethnobiologist seems to be arising, one that is less specialized, less
> politically naive, and more conscious of his or her social role. Two
> factors have played a critical role in this metamorphosis: the participa-
> tion of these new ethnobiologists in multi-disciplinary research groups,
> and their recognition that the Indian groups with which they work are
> the most exploited and marginal sector in Mexican society.

We can be sure that a fully mature ethnobiology has emerged, suggests Chilean cultural rights activist Camila Montecinos, when ethnobiological studies are no longer dominated by those "outside" the cultures of concern but are directed by those "inside" the communities (Toledo 1995). Following Montecinos and her Mesoamerican colleague Narciso Barrera, I hope for the day when ethnobiological field studies will be accomplished largely by native speakers immersed in their own cultural traditions of biological knowledge, selectively integrating other tools for learning from cultural ecology, conservation biology, and linguistics, in order to serve the needs of their communities. In the meantime, I am devoting much of my current field efforts to training young Comcáac adults in the field methods of ethnobiology and conservation biology, helping them obtain certification for the skills and knowledge they have acquired from both traditional and Western scientific practitioners.

AT THIS POINT, you might be rolling your eyes and moaning, wondering how an outsider can pretend to make sense of the many ways the Comcáac interact with the wild world around them. Very few non-Seri converse comfortably in Cmique Iitom, the extant Comcáac tongue—one distant from other living languages—and I am not one of them. As neither a speaker nor a native son, therefore, I go into this ethnobiological game with a serious handicap. Another handicap I face is the sheer biological uniqueness of the Sea of Cortés, its islands, and its coastal communities. The deserts and seas stretching out from the two Comcáac villages are home to one of the richest and most bizarre faunas of reptiles in the Americas, one that herpetologists are only beginning to understand. I was trained as an ethnobiologist and plant ecologist, not as a herpetologist, linguist, or folklorist, so any insights I have to offer will be necessarily skewed.

I am giving you fair warning: if you believe that an outsider can offer you the definitive account of another culture and its diverse interactions with wildlife, you've come to the wrong person. I concede that I cannot offer you such a thing. What I can offer you, though, are some vivid impressions of how the Comcáac distinguish the reptiles around them, how they decide which can be used and which are taboo. I can tell you what they have told me of these reptiles' interactions with other animals and with plants. And I can suggest to you why such knowledge should matter to the rest of us.

In Mr. Palomar (1983), Italo Calvino observes that "new knowledge . . . does not compensate for the knowledge spread only by direct oral transmission, which, once lost, can never be regained or transmitted." To gain a sense of how such orally transmitted knowledge is structured and embedded within specific Comcáac cultural forms, I have collaborated with ethnomusicologists in recording Xepe an Cöicóos, the songs of sea creatures; with cultural educators conversant in the ways traditional knowledge is currently being passed down to Indian children; and with linguists concerned with cognitive and linguistic acculturation in cultural communities. To understand just how unique Comcáac knowledge of local natural history may be, I have worked with zoologists familiar with the habits and habitats of desert and marine reptiles and with biogeographers sorting out how the Comcáac have influenced faunal composition of the islands of the Sea of Cortés.

I first visited the Comcáac homelands in February 1971 when I stopped in Punta Chueca during a kayak trip. As I made my way from the mangrove lagoon of Estero Santa Rosa to Isla Tiburón for a day hike and a night camping amid howling coyotes, I had my first halting conversations with these indigenous artisans and seafarers. Later in the 1970s I sporadically visited Punta Chueca and Desemboque, both as a Prescott College student of marine ecology and as a collector of cactus fruit and mesquite pods for a research project on Seri ethnobotany sponsored by the Arizona-Sonora Desert Museum (Nabhan and Mirocha 1985). I joked and traded with Comcáac families on these trips, and read everything that my mentors Richard Felger and Becky Moser wrote about them.

Curiously, it was during another period of my life, after a decade away from the Sea of Cortés coast, that I became more deeply engaged with Comcáac artisans and the conservation issues they faced. Around 1990 I was living on the U.S.-Mexico border in Organ Pipe Cactus National Monument, when loads of "Seri" ironwood carvings of turtles, snakes, and lizards suddenly began appearing in bordertown marketplaces. At about the same time, I noticed that ironwood trees were disappearing from my cactus study plots on U.S. lands managed by the National Park Service. When I asked a woodcutter in the border town of Sonoyta, Sonora, what was going on, he laughed grimly and said that hundreds of non-Indians had become "instant Seri" in order to market their animal figurines to gullible American tourists. The demand was so great, he added, that the price of ironwood had doubled over the previous year (Nabhan and Klett 1995).

Jim "Santiago Loco" Hills, an old friend who in the 1970s had made his living trading Seri carvings, convinced me to approach the Comcáac community leaders about this problem. They were enraged by the machine-made carvings imitating their own and by the deforestation of ancient ironwood stands, but felt they could do little to set matters right on their own. Santiago and I proposed to them that an ironwood conservation consortium, La Alianza Pro–Palo Fierro, be organized, to attempt to curtail the exploitation of their traditional resource and to distinguish in the marketplace authentic hand-crafted Seri animal carvings from machine-made imitations (Nabhan and Carr 1994). Within months, more than seventy Seri individuals had signed on as co-sponsors of the alliance, which was eventually successful in getting the Mexican government to confer ironwood special protection status and to recognize the Comcáac community's intellectual property rights to the animal carving tradition.

During the years I spent working with the Comcáac to protect ironwood and their rights to it, they gradually shifted to carving more with local stone than with ironwood, in part to distinguish their own artistry from that of their competitors. Whenever an artisan would ask me or Santiago how to give their handicrafts a distinctive niche in the Indian arts market, we would say that they should focus on representing the animals that are most richly celebrated in their other cultural traditions. By watching what they then selected as subjects to carve, I came to realize just how deep their appreciation of desert and marine reptiles truly was.

Soon I received carvings of all kinds of sea turtles, land tortoises, chuckwallas, and rattlesnakes, including some eating horned lizards. The carvers recounted to me elements of their creation stories about these animals, and commented on reptilian interactions with other kinds of wildlife. They nicknamed me *Hant Coáaxoj,* "Horned Lizard," and taught me how to sing the Comcáac nursery rhyme about this creature. I learned that they knew how to sex iguanas and tortoises, knew where to seek out these animals in special habitats, and regarded several species of reptiles highly for their supernatural powers. At some point, I realized that I could not truly understand Comcáac culture without factoring in their relationship with the herpetofauna of the Sea of Cortés coastline and islands. The two were inextricably linked.

As my old friend Helga Teiwes, a world-renowned ethnographer, began to photograph the Comcáac reptile carvings I brought back to the Desert Museum, I conceived of a small book celebrating the "Seri crafts revival" and the Comcáac community's involvement in wildlife conservation. It would be illustrated with images of the Seri animal carvings, and the text would include excerpts of Comcáac legends and songs. I proposed this to the Comcáac tribal chairman, the late Pedro Romero, who approved the project, as long as derivative materials from the research made their way into the Seri schools through culturally adapted educational pamphlets. Together with several talented educators, we turned our attention to this task. Then, as I accumulated hundreds of pages of notes on Comcáac knowledge of desert and marine wildlife, the concept for a promotional booklet gradually shifted and broadened.

In February 1998, Tribal Chairman Ignacio "Nacho" Barnet offered me written approval of this book project as part of an integrated, collaborative effort to promote not only Seri crafts but also the rights of the Comcáac people to use and manage reptilian resources throughout their historic territory, both within and beyond their presently decreed reservation lands. More recently, Tribal Governor Moisés Méndez and Council of Elders Chairperson Antonio Robles signed an agreement with me that not only approves and endorses this book but also establishes means by which any post-publication royalties can be used for the Traditional School recently initiated in Punta Chueca.

What you hold in your hands is my modest attempt to honor the entire range of Comcáac responses to reptilian life—naturalistic, aesthetic, mythic, and utilitarian. This is not, however, the same book that a Seri para-ecologist might write, for it reflects the relationship between the Comcáac community and the conservation biology community as much as it elucidates the Comcáac community and its surrounding faunal community of reptiles. It echoes a triangle of ecological relationships, because I have offered commentaries from conservation biologists when they seem pertinent, even though they may not be corroborated by Seri para-ecologists. Although I am the sole writer of this book, I consider my role in this work to be more that of an editor, or even a weaver, of the many voices that have contributed to this effort.

Whenever newcomers visit the Comcáac villages with me, they tend to ask me questions I can barely answer, despite a quarter century of bantering, bartering, traveling, and travailing with Comcáac community members. For example: As seafaring hunting and fishing folk, are the Comcáac mercilessly cruel to animals, or did they formerly live in some balance with the creatures around them? Is recording their knowledge of reptiles merely a nostalgic exercise in documenting an obsolete means of living in hyperarid coastal habitats, or does it celebrate universal truths with which modern societies need to be reacquainted? By dwelling on Comcáac consumption of snakes and lizards—among the "most primitive" of land animals—am I not reinforcing racist or romantic notions that the Comcáac themselves are "primitive"? Does Comcáac knowledge of wildlife extend beyond what Western-trained scientists have accumulated for the same species? Is that knowledge consistent with or antithetical to formal scientific precepts?

It is not that these questions do not interest me; they do. It is merely that other concerns have moved before them in my head and in my heart.

I have sat in rapture listening to Comcáac creation stories about nearly extinct Leatherback Turtles. I have felt my heart soar as we stalked, chased, and captured the Piebald Chuckwallas and Spiny-tailed Iguanas of Isla San Esteban. Once during a fit of anger and despair over difficult interpersonal relationships I was calmed by the lovely performance of a Western Whiptail song by an old blind woman named Eva, who has since passed away. I have been intellectually and spiritually consumed by the questions of why pregnant Seri women should not be allowed to view banded geckoes, and why the bearers of boomboxes, CB radios, and Timex watches still claim that Desert Tortoises and Leatherback Turtles speak to them. I have learned that sea turtle hunters sing constantly while they are out in their boats, keeping their minds focused on images of turtle behaviors that can subconciously alert them to the fleeting appearance of a turtle out of the corner of their eye.

I have felt amazement, frustration, and remorse as I watched the Comcáac community attempt to hang on to its traditions despite the many pressures moving against it. And I have laughed myself to tears during some of our cross-cultural encounters, when it became hilariously obvious to both my Comcáac friends and myself that we do not at all see, feel, hear, or smell the world in the same way. If I have learned anything, perhaps it is that we are sublimely enriched by this diversity among neighbors, just as we are by the reptile community around us.

THIS BOOK is divided into two mutually reinforcing parts. Part 1 is a narrative account intended for any reader interested in traditional ecological knowledge. Part 2 is in essence a status report on what is known of the herpetofauna of the Comcáac homelands, from both the indigenous and Western scientific perspectives.

Traditionally, ethnobiological monographs present species-by-species accounts of knowledge specific to a particular culture. While many such studies also include excellent introductory overviews of the people, their history, and the landscape, they do not necessarily treat topics such as island biogeography and conservation ethics currently being debated by scholars and activists.

In this book, instead of describing the Seri and their environment from my perspective alone, I have attempted to use the Seri people's voices as well to establish their place in these global debates. The two parts, therefore, are complementary: part 1 raises issues and offers general answers, whereas part 2 provides the details that may allow others to judge whether I have correctly interpreted Seri ethno-systematics and cultural and economic practices.

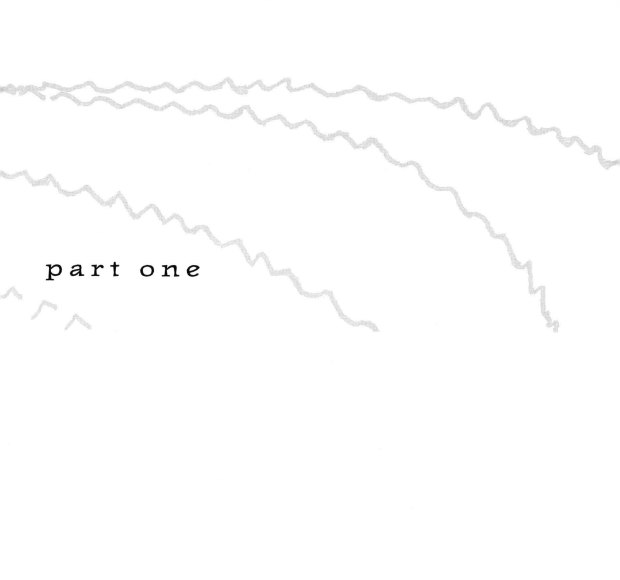

part one

islands of uniqueness

Endangered Cultural Knowledge of Endemic Creatures

Both species and languages have evolved over hundreds or thousands of years to adapt to very specific environmental and sociopolitical contexts. If those contexts undergo unprecedented rapid change—as the world's environments and cultures are now doing—many species and languages will likely lack the resiliency to adapt to new conditions. In biology, island-dwelling species are hallmarks of such highly specialized, highly vulnerable lifeforms; and sure enough, exactly 75% of all recorded animal extinctions occurring since 1600 have been of island species.

DAVID HARMON, "The Converging Extinction Crisis" (1996)

Our nostrils flared as we entered the smoky plume of the fiesta, walking
into the gathering from the other side of the village's largest satellite dish.
The acrid bite of woodsmoke from elephant tree and mesquite filled the
sandy opening between the rows of houses on the northern edge of Punta
Chueca. Because we arrived in the dark, we had no idea how the fiesta was
staged or where we should stand. The meager light projecting out from
nearby houses and from a lone cooking fire was barely enough to guide
us between the circles of singing, gambling, dancing, and eating.

I told my friends to go on ahead while I let my eyes adjust. I took a deep
breath and tried to figure out what I was doing out at midnight at a Comcáac puberty ceremony on the shores of the Sea of Cortés.

I could smell and hear meat sizzling on a grill hidden somewhere beyond
my view. I meandered through the crowd, trying to find where food was
being served, and ended up beneath a newly constructed ceremonial bower
where the girl of honor stood. The bower of freshly cut branches inundated
the puberty celebrant in the rich fragrances of ocotillo, red elephant tree,
and desert lavender, but I could not identify them by sight at the time, the
light was so spotty. A broth was cooling in a big pot nearby; I located it by

following its vapor trail and savory smell. My vision blinked on and off as Comcáac fiesta-goers walked back and forth in front of two weak spotlights and the cookfire. At the same time, my nose took in a steady flow of aromatic information.

Earlier that afternoon, while we were working on constructing a sanctuary for threatened chuckwallas, the tribal chairman's wife had come by to invite us to the celebration marking her niece's passage into womanhood. Finishing up work after dark, we then went searching for the fiesta, and that is when we heard the partying on the other side of the satellite dish. The honoree looked no more than a girl, one perhaps prone to listen more to Latin rock on the radio than to her uncle's Yaqui-style *pascola* songs being sung a few feet away from her. She might have wanted to be at a Mexican-style *quinceañera* party where she could dance the *cumbia* with a prospective boyfriend, instead of letting the world know that she had recently made her passage into womanhood.

And yet there she was, giggling and whispering with her sister while boys her age danced to the *pascola* songs one by one, shaking the silk-moth cocoon rattles on their ankles, making their feet sing like those of flamenco dancers. She must have known how much the persistence of this ceremony meant to the elders around her, including her grand-parents. By acquiescing to be the honoree, she was allowing her entire Comcáac community to maintain its spiritual traditions even as massive social, economic, and environmental changes were churning up all around them. As we kept an eye on her, one of her young, slightly inebriated cousins kept his eye on us. He would approach us and very seriously request that we sing an American song or recite our life histories into his tape recorder, so that he could "save" this event for posterity.

While most of my co-workers tried to shake this amateur anthropologist by inching into the crowd formed around the *pascola* dancers, my partner Laurie and I sat and tried our hand at the traditional gambling games. I teamed up with a former tribal chairman in the men's game of sleight of hand: "Which cane reed is filled with sand?" We were great at baffling the others, but poor at reading their poker faces when it came time for them to fool us. Laurie had better luck, winning several necklaces from women who sat around a circle made from slices of Organpipe cactus stems. Soon Laurie was invited by the honoree's mother to help herself to some of the feast food being served on the other side of the campfire.

A minute later, Laurie nudged me and whispered, "They're serving tacos made with Green Sea Turtle meat." She handed me a plateful.

"¿Caguama? ¿Carrinegra?" I asked, a bit incredulous.

Sea turtles had declined so precipitously in the Sea of Cortés that I had seen only a handful of their carapaces in Punta Chueca over the previous year, and those were from turtles accidentally caught in fishing nets. Two decades before, freshly butchered carapaces were ubiquitous, so much so that they were frequently used as roof thatch on traditional huts. The Mexican government finally placed a ban on commercial sales of all sea turtle products, but the Comcáac still have rights to capture Green Sea Turtles for ceremonial purposes.

"¿Qué tipo de moosni es?" I asked Ernesto Molina, who stood beside us nibbling on a turtle taco.

"Cooyam," he uttered between mouthfuls, referring to the younger migratory Green Sea Turtles that arrive from the south in the spring earlier than the rest.

Ernesto finished his taco, then asked us, "Have you tried their juice—the sea turtle broth?" He disappeared into the crowd and, when he returned, handed me a cupful of turtle broth. It tasted a bit like one of my favorite Mexican dishes, birria. The sweet broth was still cooling, and as it did, tender, oil-rich flakes of turtle meat floated to the top.

As I tossed my head back to down the last spoonfuls of broth, I saw for the first time what had been above us the entire evening. Tied to a branch directly above the men's gambling circle was the freshly butchered head of a sea turtle, dark green scales glistening in the meager light.

It dawned on me that we were participating in a sacrament, one that has been performed ever since the Comcáac first became seafaring people. By blessing this young woman's rite of passage with the meat and blood of *moosni cooyam*, they were linking her life to the very creature that swims through their culture's stories, songs, dreams, and diet.

I felt honored and humbled to be part of the communion. But as I looked up again into the face of the sea turtle shining in the firelight, another wave of emotion washed over me. Because I once shared quarters with a marine biologist who worked tirelessly to protect the nesting beaches of sea turtles, I had for twenty years boycotted any restaurants that featured sea turtle meat, eggs, or soup.

Caught up in the moment, perhaps flattered by the invitation to share sea turtle with Seri friends, had I suffered some ethical lapse, somehow forgetting that sea turtles are endangered? Or had I not let my ethics slip, but instead accepted a tenuous balance between how I express my concern for an endangered people and how I express my concern for an endangered animal?

The young Seri anthropologist reappeared with his tape recorder, which no longer held either tape or batteries. Nevertheless, he wanted to talk with me. He held the recorder up in front of my mouth.

"Well," he began, "tell me about your culture. What are your beliefs?"

A GRILLED MORSEL of turtle meat passes between hands, then disappears between two sunburned lips. Sacramental meals of sea turtle have been shared on the shores of the Sea of Cortés for centuries, especially among those seafarers known to others as the Seri. While those Comcáac families living along the central Gulf coast of Sonora have always savored sea turtle meat, they especially value the opportunity to participate in this ritual today, because it takes place less and less frequently. Turtle meat has become scarce.

It is a scarcity that would have been unimaginable to Comcáac seafarers a century ago. The five sea turtle species formerly abundant off the west coast of Mexico have recently been decimated by the loss of nesting habitat, incidental

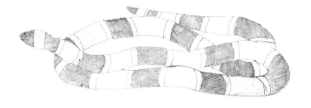

Figure 1. A Western Coralsnake, known to the Comcáac as *coftj*, is considered both poisonous and dangerous by many biologists, but not by most Seri individuals. This drawing by Gabriel Hoeffer of Desemboque was commissioned by Gary Nabhan in 1998.

take by shrimp trawlers and drift nets, overharvesting by commercial turtle fishermen, and clandestine collection of turtle eggs on beaches. Each time I see a Seri friend partaking of this traditional food, I wonder how long it will be before such consumption is no longer permitted by federal or international law, or no longer possible as a result of the complete collapse of sea turtle populations. Comcáac children today seldom see the roasting or smell the sizzling of turtle meat, even though their grandparents grew up immersed in those sights, tastes, and smells.

Despite these declines, however, sea turtles and many other reptiles remain present in the region; fortunately, they also remain present in the songs, stories, and art of indigenous residents of the Sonoran coast (Fig. 1). Whenever I have gone with Comcáac fishermen to the islands of the Sea of Cortés, they have shown me turtles, tortoises, lizards, and snakes that I would be unable to see anywhere else in the world (Fig. 2). From fishermen and herbalists I have heard commentaries on natural history, local uses, and appropriate ways of harvesting that cannot be heard in villages of other cultures nearby.

Figure 2. Alfredo López displays a freshly captured Piebald Chuckwalla on Isla San Esteban. He is one of several middle-aged men who now work part of the year as guides and collaborators for wildlife biologists in the midriff islands. (Photo Gary Nabhan, 1998)

"There are rattlesnakes without fully formed rattles on Islas San Esteban, San Lorenzo, and Ángel de la Guarda. And on Tiburón, there are Desert Tortoises that for us are like the Leatherback Sea Turtles, [in the sense that] they could talk with us because they were once people themselves. The Piebald Chuckwallas of San Esteban? We know how to find them, to get them to move, by making a windlike noise with a bullroarer."

ERNESTO MOLINA, *Punta Chueca*

I believe this unique body of cultural knowledge is worth paying attention to, as much as I believe that the endemic plants and animals of this coastal desert warrant our curiosity, care, and delight. Even if you or I never eat a Black Chuckwalla, harm a Regal Horned Lizard, or use a local saltbush to treat snakebite, the ways in which the Comcáac community engages itself with these plants and animals is worthy of our understanding.

Figure 3. The central Gulf coast region of the Sonoran Desert, far from being a barren "wasteland," supports some of the highest levels of endemism among native plants and animals of any landscape in North America. (Photo Gary Nabhan, 1997)

Much of this ethnobiological knowledge can be expressed only in Cmique Iitom, the Comcáac tongue, which is considered a language isolate, unrelated to any other language now spoken in Sonora, or anywhere else in the world for that matter (Marlett 2000). It is also a body of knowledge restricted to the driest portions of the Sonoran Desert and the midriff islands in the Sea of Cortés (Fig. 3). That coastal desert, with the stunning beauty of its seascapes and rugged mountains, has a bewildering capacity to support a diversity of life-forms seldom seen anywhere else in such odd juxtapositions.

"Before, the boojums [an endemic plant] were people—people named Cototaj. They were once people who were trying to climb high up to the tops of those hills when it [the Great Flood] happened. They were climbing because they were terrified of the tide that was rising. In that time, tidal waves came to terminate the world. This tide was rising up toward the top of the world, to finish off the world. The Cototaj were so scared that they were trying to escape. They were people then, but they were changed [into giant plants] as the rising sea reached their feet."

ADOLFO BURGOS OF DESEMBOQUE, *visiting the only boojum population in Sonora*

To place such cultural knowledge about wildlife and landscapes in perspective, it is worth remembering that the Comcáac are engaged daily with life-forms and land-forms not necessarily found in other parts of the world. Few members of other cultures, and even few natural scientists, have spent as much time observing certain unique reptiles and plants in their habitats as the Comcáac have. Mexico as a whole, and the Sea of Cortés region in particular, are rich in what biogeographer Eduardo Rapoport (1982) calls *micro-areal endemics*. However clumsy this term, it is all we have to refer to plants or animals limited to small geographic areas, with distributions covering no more than a 250-kilometer by 100-kilometer extent—an area so small that a Comcáac traveler could circumscribe their entire range within a week's walk. The Sonoran Desert has many species of plants, as well as some mammals and reptiles, with such small ranges.

Although no precise estimate exists for the number of micro-areal endemics within Mexico, fair estimates have been made of the number of species that are *geopolitically endemic*, that is, restricted entirely within the Republic of Mexico, or *bioregionally endemic*, found within bioregions that are centered in Mexico. The Republic of Mexico ranks second among all nations for harboring the greatest number of vertebrate animal species—at least 761—that live nowhere else in the world (Harmon 1995). It ranks fourth among nations for the size of its flora, which encompasses some 22,000 vascular plant species (Rzedowski 1993). Within "MegaMéxico 1," a bioregion defined by biogeographer Jerzy Rzedowski (1993) as Mexico plus the U.S. Southwest, desert scrublands and semi-arid grasslands host the greatest number of endemic species—at least 5,600—of any biome or habitat complex.

Significantly, Mexico is home to 230 languages unspoken beyond its boundaries, placing it sixth among nations in terms of extant cultural diversity and endemic languages (Harmon 1995). Because most unique cultural knowledge about plants and animals is "encoded" in specific terms for describing behavior, morphology, and habitat, assessments of linguistic diversity are often the first steps taken as scholars attempt to gauge a region's specialized ethnobiological knowledge.

Although the arid scrublands of Baja California and the Chihuahuan Desert region have lost most of the ethnic groups that lived there during the pre-Columbian and colonial eras, Sonora's indigenous desert dwellers have fared far better. Most students of Mexico's cultural history recognize the contributions made by the ethnic groups commonly known as the Yaqui and Mayo (Yoemem), the Seri (Comcáac), the Cocopa (Cucupá), and the Papago-Pima (O'odham), who collectively number in the tens of thousands today. The populations of most of these groups are larger than a century ago, though that does not mean that more people now speak their native languages. Nor does it mean that cultural knowledge about local plants and animals remains intact, for not all residents of the region have exposure to the plants and animals endemic to the desert and sea.

The Sea of Cortés region remains a prominent contributor of endemics to the biota of Mexico as a whole, though the Seri, Yaqui, Mayo, and Cocopa are the only indigenous groups still making a living along its shores. The islands of the Sea of Cortés, as well as coastal Baja California and coastal Sonora, are noted for their impressive roster of endemic fauna and flora (Tables 1 and 2). Of plant species unique to the region, 42.8 percent are known, used, or named by the Comcáac community (Felger and Moser 1985; Nabhan field notes). I have searched ethnobiological and linguistic literature for neighboring cultures' knowledge of these same organisms and have yet to find even 15 percent of the species on these lists that are named in other languages. Of course, names are merely the entry point into the domain of cultural knowledge that a community may share. And yet, if one culture does not even name an organism, while another not only names it but also has terms to describe its morphology or anatomy, it is likely that the latter culture has unique knowledge of the ecology and behavior of the organism in question. The Comcáac do encode cultural knowledge about certain geographically restricted species in their language—knowledge about where these organisms live, when they reproduce, and what they are "good for." This is knowledge that would likely not be found among neighboring communities if the Comcáac themselves ever "lost" or "abandoned" it, for whatever reasons.

Could it be that after centuries of intermittent contact with O'odham, Yoemem, Cucupá, and mestizo cultures, the Comcáac have retained a knowledge of the natural world that is marked by limited borrowing, and hence unique? I have come across only a few O'odham and Yoemem uses of the Comcáac term for sea turtles (moosni), and rare Comcáac appropriations of O'odham terms, such as for broomrape (matar, from mo'ostalk) and for bighorn (ceso, from ceşoiñ) (Felger and Moser 1985). It seems that when the Comcáac engaged in hunting and fishing over the last few centuries, they discussed wildlife largely in their own language, not in Spanish—the lingua franca for other economic activities—or in a neighboring Uto-Aztecan dialect.

TABLE 1

Endemic Terrestrial Fauna of Large Midriff Islands Visited by Comcáac Seafarers

Species	Island
LIZARDS	
CROTAPHYTIDAE	
Crotaphytus insularis	Ángel de la Guarda
IGUANIDAE	
Ctenosaura conspicuosa	Cholludo, San Esteban
C. nolascensis	San Pedro Nolasco
Dipsosaurus catalinensis	Santa Catalina
Sauromalus hispidus	Many
S. klauberi	Santa Catalina
S. slevini	Carmen, Coronados, Monserrate
S. varius	Roca Lobos, San Esteban
PHRYNOSOMATIDAE	
Petrosaurus slevini	Ángel de la Guarda, Mejía
Sceloporus angustus	San Diego, Santa Cruz
S. grandaevus	Cerralvo
S. lineatulus	Santa Catalina
Uta lowei	El Muero
U. nolascensis	San Pedro Nolasco
U. palmeri	San Pedro Mártir
U. squamata	Santa Catalina
U. tumidarostra	Coloradito
EUBLEPHARIDAE	
Coleonyx gypsicolus	San Marcos
GEKKONIDAE	
Phyllodactylus bugastrolepis	Santa Catalina
P. partidus	Cardonosa Este, Partida Norte
TEIIDAE	
Cnemidophorus bacatus	San Pedro Nolasco
C. canus	Salsipudes, San Lorenzo Norte and Sur
C. carmenensis	Carmen
C. catalinensis	Santa Catalina
C ceralbensis	Cerralvo
C. danheimae	San José
C. espiritensis	Espíritu Santo, Partida Sur
C. franciscensis	San Francisco
C. martyris	San Pedro Mártir
C. pictus	Monserrate

TABLE I (continued)

Species	Island
SNAKES	
COLUBRIDAE	
Chilomeniscus punctatissimus	Espíritu Santo, Partida Sur
C. savagei	Cerralvo
Eridiphas marcosensis	San Marcos
Hypsiglena gularis	Partida Norte
Lampropeltis catalinensis	Catalina
Masticophis barbouri	San Pedro Mártir
M. slevini	San Esteban
Rhinocheilus etheridgei	Cerralvo
VIPERIDAE	
Crotalus angelensis	Ángel de la Guarda
C. catalinensis	Santa Catalina
C. estebanensis	San Esteban
C. mitchellii	Espíritu Santo, Partida Sur, Salsipuedes, others
C. muertensis	El Muerto
C. tortugensis	Tortuga
MAMMALS	
CANIDAE	
Canis latrans jamesi	Tiburón
CERVIDAE	
Odocoileus hemionus peninsula	San José
O. h. sheldoni	Santa Cruz
HETEROMYIDAE	
Dipodomys insularis	San José
D. merriami mitchellii	Tiburón
Perognathus arenarius siccus	Cerralvo
P. baileyi insularis	Tiburón
P. penicillatus seri	Tiburón
P. spinatus bryanti	San José
P. s. evermani	Mejía, Granito
P. s. guardiae	Ángel de la Guarda, Las Ánimas
P. s. lambi	Espíritu Santo
P. s. latijugularis	San Francisco
P. s. marcosensis	Santa Catalina
P. s. occultus	Carmen
P. s. pullus	Coronados
P. s. seorsus	Danzante

Continued on next page

TABLE I (continued)

Species	Island
LEPORIDAE	
Lepus alleni tiburonensis	Tiburón
L. californicus sheldoni	Carmen
L. insularis	Espíritu Santo
Sylvilagus mansuetas	San José
MURIDAE	
Neotoma albigula seri	Tiburón
N. bunkeri	Coronados
N. lepida abbreviata	San Francisco
N. l. insularis	Ángel de la Guarda, Mejía
N. l. latirostra	Espíritu Santo
N. l. marcosensis	Santa Catalina
N. l. nudicaula	Carmen
N. l. perpallida	San José
N. l. vicina	Espíritu Santo
N. varia	Dátil
Peromyscus caniceps	Monserrate
P. dickeyi	Tortuga
P. ermicus angelensis	Ángel de la Guarda
P. e. avius	Cerralvo
P. e. carmeni	Carmen
P. e. cinereus	San José
P. guardia guardia	Ángel de la Guarda
P. g. mejiae	Mejía, Granito
P. interparietalis	Las Ánimas, Salsipudes, San Lorenzo
P. pseudocrinatus	Cerralvo
P. sejugis	San Diego, Santa Catalina
P. slevini	San Marcos
P. stephani	San Esteban
PROCYONIDAE	
Bassariscus astutus insulicola	San José
B. a. saxicola	Espíritu Santo
SCIURIDAE	
Ammospermophilus insularis	Espíritu Santo
VESPERTILIONIDAE	
Myotis vivesi	Tiburón

Sources: Grismer 1999; Case and Cody 1983; Figueroa 1999.

TABLE 2

Endemic Terrestrial Flora of Large Midriff Islands and Adjacent Central Gulf Coast

Species	Range	Used or Named by Comcáac
AGAVACEAE		
Agave cerulata var. dentiens	San Esteban	Used as food
A. felgeri	Bahía San Carlos, Sonora	Probably used as soap
A. pelona	Sonoran coast	Used as food
A. subsimplex	Sonoran coast	Used as food and in art
Brahea armata	Ángel de la Guarda, Tiburón, Baja coast	Known, but not used
B. roezlii	Bahía San Carlos, Sonora	Used as headpieces
ASTERACEAE		
Driopetalum crenatum var. racemosum	Espíritu Santo	No
BORAGINACEAE		
Cryptantha grayi var. racemosum	Espíritu Santo	No
CACTACEAE		
Echinocereus grandis	San Esteban, San Lorenzo	Used as food
E. websterianus	San Pedro Nolasco	No; related species used
Ferocactus diguetii	Ángel de la Guarda	No; related species used
FABACEAE		
Acacia willardiana	Tiburón, San Pedro Nolasco, and Sonoran coast	Used for tools and shelter
POLYGONACEAE		
Eriogonum angelense	Ángel de la Guarda	No; related species used

"During the era of Coyote Iguana [a Seri who kidnapped and married a Mexican girl, Lola Casanova, around 1850], a full-blooded Seri woman we call Virginia lived at Pozo Coyote. Her daughters were among the few who married Papago, Yaqui, and Spanish men who would come there to trade skins of deer, javelina, and cloth. But it is an old Seri village, one we call Hatájc Ano Ziix Coocö—Where Coyote Came to Drink."

AMALIA ASTORGA OF DESEMBOQUE, *while visiting Pozo Coyote*

Except through trading, a few intermarriages, and a few joint fishing and foraging trips where their territories overlap with those of other indigenous groups, the Comcáac historically had little chance to share their unique ethnobiological knowledge with other cultures, a sharing that occurred with Nahuatl and Cahitan speakers. From my perusal of Sobarzo's *Vocabulario sonorense* (1966) and Molina's *Nombres indígenas de Sonora y su traducción al español* (1972)—two linguistic compendia of loan words from the region's indigenous languages—it is clear that plant and animal names from Cmique Iitom have entered into the regional lexicon far less than those from other native languages. It is therefore appropriate to speak of their understanding of the native biota of the central Gulf coast of Sonora and the midriff islands nearby as *endemic ecological knowledge*.

David Harmon, a co-founder of Terralingua, a group concerned with the loss of native languages, offered one of the first perspectives on the importance of this endemic ecological knowledge:

> Endemicity calls to mind another intriguing possibility: that numerous small cultural groups have co-evolved with the locally adapted animals and plants around them. While this possibility must remain quite speculative at this point, the intimate knowledge indigenous peoples often have of their immediate environment, and their traditional dependence on it, beckons us toward further consideration of co-evolution. Whatever the case, traditional environmental knowledge is increasingly being seen as valid scientific information by biologists. . . .
>
> Unfortunately, as more and more native groups are set adrift in the flood tide of globalized pop culture, this knowledge runs the risk of being lost. (Harmon 1996)

"When I was tribal governor a few years back, I realized that we faced more subtle threats [than genocide]. Tourism promoters wanted to build hotels and roads on our islands, saying that it would benefit the Seri people. 'Sign here,' they'd say, 'it's simple.' But I wouldn't sign off on any development on the islands. Finally, a politician got mad at me and said, 'Just how much do you think your *pinché* island will ever be worth if it isn't developed?'

"That made me mad. I said to him, 'That island is worth more than you could ever pay us, because it has had the blood of my people's veins running into its sand. It is offensive to us to think of selling land that our people have died for.'"

GENARO HERRERA, *former tribal governor, Punta Chueca*

When *Time* magazine presented its cover story "Lost Tribes, Lost Knowledge" by Eugene Linden (1991), much of the coverage dealt with the unique ecological and medical knowledge encoded in endemic languages. Linden was preoccupied with "rainforest peoples" who restrict their movements to relatively small areas of the Amazon Basin, know the local wildlife there intimately, and now suffer from the impacts of deforestation. He is, moreover, one of many observers to relate the degree of geographic restriction of various indigenous populations to their vulnerability to extinction by destruction of their habitat. As a humanitarian, Linden argued that these geographically restricted peoples need support simply because their younger generations deserve a future in their fatherlands. Yet he also argued that we should be concerned with their plight as well because they hold unique knowledge about medicinal plants and other natural resources that potentially offer benefits to other societies. If these "endemic" tribes, their languages, or their knowledge becomes lost, the rest of us will be unable to benefit from the cultural and ecological insights they accumulated over thousands of years.

Of course, saying that a creature, a language, or a cultural knowledge base is endemic is not the same as saying that it is endangered. Designating any entity as endangered—whether indigenous culture or native creature—is not merely to conclude that its continued decline is highly probable; it is also to make an emotionally charged and politically mediated judgment.

"Many people have said that indigenous peoples are myths of the past, ruins that have died. But indigenous communities are not vestiges of the past, nor are they myths. They are full of vitality and have a course and a future . . . [with] much wisdom and richness to contribute. They have not killed us and they will not kill us now. We are stepping forth to say, 'No, we are here. We live.' "

RIGOBERTA MENCHÚ, *Nobel Peace Prize laureate*

Too often, the pronouncement of "endangerment" sounds like an obituary rather than an early warning that additional protective or supportive measures should be put into place. This is why activists involved in ecological restoration and language revitalization efforts do not believe that offering support only to those closest to extinction is sufficient to safeguard diversity; they emphasize the need to safeguard entire communities rather than just a few members of those communities.

Despite the efforts of scholars such as Mace and Lande (1993), Soulé (1987), and Harmon (1996) to propose more comprehensive criteria for determining when a species or language is threatened with extinction, such criteria have yet to help those in the field. In particular, field biologists and resource managers have been forced into a "triage" mindset, having to recommend to environmental policymakers which species deserve immediate attention even though they often lack quantitative data documenting rates of decline. When biologists noticed that certain species endemic to the Sea of Cortés region had become exposed to new collecting threats, for instance, they urged Mexican government agencies to provide official protection for them. However, it was later learned that threats from collectors had far less impact on these species than did invasive species and habitat degradation. At the least, such recommendations have provided incentives for better monitoring the changing status of these species of concern.

The central Gulf coast of Sonora and the adjacent midriff islands are recognized as home to no less than twenty species of reptiles listed in Mexico's Di-

ario oficial as threatened, endangered, or deserving of special protection status (Table 3). An additional six marine species that venture into the midriff area are also listed as globally threatened (Table 4), for example, the Leatherback Sea Turtle (Spotila et al. 1996).

TABLE 3

Terrestrial Species Considered Endangered, Threatened,
or Deserving of Special Protection Status in the Central Gulf Coast

Species	English Name	Spanish Name
PLANTS		
AGAVACEAE		
Agave felgeri	Felger's Amole	Amole
A. pelona	Bald Mescal	Mescal Pelón
ARECACEAE (PALMAE)		
Brahea armata	Mexican Blue Palm	Palma taco azul
B. elegans	Nacapuli Palm	Palma de Nacapuli
CACTACEAE		
Ferocactus cylindraceus	Cylindrical Barrel Cactus	Biznaga
F. johnstonianus	Island Barrel Cactus	Biznaga
Mammillaris angelensis	Isla San Esteban Fishhook Cactus	Cabeza del viejo
M. estebanensis	Isla San Esteban Fishhook Cactus	Cabeza del viejo
Peniocereus striatus	Sonoran Queen of the Night	Zaramatraca
FABACEAE		
Olneya tesota	Ironwood	Palo fierro
FOUQUIERIACEAE		
Fouquieria columnaris	Boojum	Cirio
MALVACEAE		
Gossypium turneri	Teta Cahui Cotton	Algodon de Teta Cahui
BIRDS		
ACCIPITRIDIAE		
Accipiter cooperi	Cooper's Hawk	Gavilán de Cooper
FALCONIDAE		
Falco peregrinus	Peregrine Falcon	Halcón peregrino

Continued on next page

TABLE 3 (continued)

Species	English Name	Spanish Name
FRINGILLIDAE		
Passerculus sandwichensis	Savannah Sparrow	Gorrión de savannas
LANIIDAE		
Lanius ludovicianus	Loggerhead Shrike	Cabezóna
PANDIONIDAE		
Pandion haliaetus	Osprey	Aguila pescador
PELECANIDAE		
Pelecanus occidentalis	Brown Pelican	Alcatraz café

TURTLES AND TORTOISES

TESTUDINIDAE		
Gopherus agassizii	Desert Tortoise	Tortuga del monte

LIZARDS

CROTAPHYTIDAE		
Gambelia wislizenii	Long-nosed Leopard Lizard	Lagartija de leoparda, cachora
IGUANIDAE		
Sauromalus hispidus	Spiny (Black) Chuckwalla	Iguana negra
S. varius	Piebald (San Esteban) Chuckwalla	Iguana de Isla San Esteban
PHRYNOSOMATIDAE		
Callisaurus draconoides	Zebra-tailed Lizard	Perrito
EUBLEPHARIDAE		
Coleonyx variegatus	Western Banded Gecko	Salamanquesa, geco
HELODERMATIDAE		
Heloderma suspectum	Gila Monster	Escorpión pintado

SNAKES

BOIDAE		
Charina trivirgata	Rosy Boa	Boa rosada
COLUBRIDAE		
Chilomeniscus stramineus	Bandless Sandsnake	Culebra de los médanos, coralillo falso

TABLE 3 (continued)

Species	English Name	Spanish Name
Hypsiglena torquata	Nightsnake	Culebra de la noche
Masticophis estebanensis	Isla San Esteban Coachwhip	Chirrionera
M. flagellum	Sonoran Coachwhip	Chirrionera
ELAPIDAE		
Micruroides euryxanthus	Western Coralsnake	Coralillo
VIPERIDAE		
Crotalus atrox	Western Diamondback Rattlesnake	Víbora de cascabel
C. cerastes	Sidewinder	Cuernitos
C. estebanensis	Isla San Esteban Rattlesnake	Víbora de cascabel de Isla San Esteban
C. molossus	Black-tailed Rattlesnake	Víbora de cascabel
C. scutulatus	Mohave Rattlesnake	Víbora de cascabel
C. tigris	Tiger Rattlesnake	Cascabel del tigre
MAMMALS		
ANTILOCAPRIDAE		
Antilocapra americana sonoriensis	Sonoran Pronghorn	Berrendo sonorense
BOVIDAE		
Ovis canadensis mexicana	Desert Bighorn Sheep	Borrego cimarrón
CANIDAE		
Canis lupus	Mexican Wolf	Lobo mexicano
Vulpes macrotis	Kit Fox	Zorro del desierto
FELIDAE		
Felix onca	Jaguar	Onza, yagour
PHYLLOSTOMIDAE		
Choeronycteris mexicana	Mexican Long-tongued Bat	Murciélago trompudo
Leptonycteris curasoae	Lesser Long-nosed Bat	Murciélago hocicudo mescalero

Sources: Mexico's *Diario Oficial*, U.S. *Federal Record*, CITES Appendix One, and IUCN/Species Survival Commission publications.

TABLE 4

Threatened Marine Fauna of the Sea of Cortés

Species	English Name	Spanish Name
FISH		
SCIAENIDAE		
Totoaba macdonaldi	Totoaba	Totoaba
TURTLES AND TORTOISES		
CHELONIIDAE		
Caretta caretta	Loggerhead Turtle	Caguama cabezona, jabalina
Chelonia mydas	Green Sea Turtle	Caguama carrinegra
Eretmochelys imbricata	Hawksbill	Carey, perico
Lepidochelys olivacea	Olive Ridley	Golfina
DERMOCHELYIDAE		
Dermochelys coriacea	Leatherback	Siete filos
CROCODILES		
CROCODYLIDAE		
Crocodylus acutus	River Crocodile	Cocodrilo, caimán
MAMMALS		
PHOCOENIDAE		
Phocoena sinus	Gulf of California Harbor Porpoise	Vaquita
ESCHRICHTIIDAE		
Eschrichtius robustus	Gray Whale	Ballena gris

"The government [environmental protection agency, PROFEPA] doesn't bother us as long as we never try to sell sea turtle meat or eggs. It's for consumption here only, as part of our ceremonies.

"We're the only ones in Mexico permitted to harvest the Green Sea Turtle because we know how to do so without damaging the resource."

GENARO HERRERA, *translated by Ernesto Molina, Punta Chueca*

When little other information is available, biologists often consider a species to be endangered when fewer than five hundred reproductive individuals remain in the wild (Soulé 1987). Certain linguists are now using a parallel benchmark, however rudimentary, for declaring that a language is endangered: if fewer than one thousand individuals remain who converse with one another in a particular native language, the language is considered to be at risk for extinction (Harmon 1996). Using data derived from a global *Ethnologue* survey of 5,635 endemic languages, Harmon (1995) determined that 1,513 languages meet this criterion, and another 220 are already "extinct." Mexico is home to 26 of those endangered languages—8.5 percent of Mexico's 230 indigenous languages—and another 4 are extinct. Among those 26 is Cmique Iitom.

I don't believe many Seri individuals would accept this designation, based as it is solely on the group size of speakers, even though they might express concern that their young people are using more Spanish at the expense of conversing in Cmique Iitom. Cmique Iitom remains spoken in the vast majority of homes in Punta Chueca and Desemboque, in tribal meetings, and even in the schools, and so seems far from a "dying language" when you are in the midst of the Comcáac community. Although the majority of teachers in Punta Chueca and Desemboque offer lessons primarily in Spanish, the schoolchildren tend to answer in Cmique Iitom and are not currently chastised for doing so. Since the use of Cmique Iitom is by no means restricted to discussions among grandparents and other elders, it would be difficult to argue that it is dying out. To the contrary, if linguists who have worked with endangered languages elsewhere in the world were to visit a Comcáac village, I believe they would be heartened by the intensity and vitality of native language use they would hear there. While elders within the community worry that their grandchildren are not exercising the full range of expressiveness in Cmique Iitom, at least these youth have not entirely abandoned their native tongue.

> "The language spoken here is changing. The young people understand only some of the words, and the shorter stories we tell them. . . . I have had to take into account for myself [as a teacher] that there are many terms in the Seri language that the children hardly know, or misuse [relative to their original meaning]."
>
> PEDRO ROMERO, *former tribal governor and teacher, Punta Chueca*

On the other hand, scarcely more than 500 people left on the face of this planet speak Cmique Iitom as their first language. In a 1995 census, the 720 residents of Punta Chueca and Desemboque included only 518 individuals who identified themselves as 100 percent indigenous in parentage; even though all of these individuals consider themselves "Seri," some of them have O'odham or Yoemem ancestry as well. Another 137 individuals identified themselves as having *mestizo*, or mixed, parentage—both Indian and non-Indian—and only a few of these residents speak Cmique Iitom as their first language. Of the 65 non-Indians, or "whites," who lived in the settlements of Punta Chueca, Desemboque, Campo Egipto, and Punta Sargento in 1995, few had more than a rudimentary vocabulary in any indigenous language. In 1996 Emilia Estrada (pers. comm.) estimated that of the 650 Seri, only 550 speak their native language. Considering that not all of the children of mixed parentage have equal confidence in Spanish and Cmique Iitom, I would guess that fewer than 500 converse as frequently in their native tongue as they do in Spanish.

Given these difficulties in determining whether Cmique Iitom is truly an endangered language, what can be said about the Comcáac as an "endangered people"? Although their total population remains rather low, and some intermarriage occurs with non-Indians, the overall size of the Comcáac community has risen dramatically over the last few decades. Whereas the two permanent villages had over 650 indigenous residents in 1995, this is nearly four times the estimate of Seri individuals made by ethnologist W. J. McGee when visiting their camps at the end of the nineteenth century (Fig. 4). Once out-and-out warfare against the Comcáac terminated after 1904, and as health care and ex-

Figure 4. This posed portrait of a young Seri woman with traditional face paintings and a white pelican-skin dress was taken during the McGee expedition to Sonora in 1894. Such images established a stereotype of the Comcáac as picturesque but remote "primitives," doing little to foster true cross-cultural understanding. The photo is courtesy of Thomas Vennum of the Smithsonian Institution, who "repatriated" copies of both audiotapes and photos in the National Anthropological Archives to the Comcáac community in 1997. (Photo William Dinwiddie, 1894)

tralocal food supplies have become available, their population has greatly expanded.

While the Comcáac seem to be recovering as an ethnic population, it is unclear whether demographic trends indicate an increasing probability of cultural persistence. In 1995, at least 56 percent of the population of the Comcáac communities was under twenty-four years of age—under the mean marriage age of twenty-six (Becky Moser, pers. comm.); only 5 percent of the population was over sixty-five years of age. By the year 2010, the number of monolingual Comcáac who grew up with virtually no daily contact with non-Indians will likely be less than 1 percent of the population. Most future culture bearers in the Comcáac community will therefore be bilingual individuals accustomed to modern technologies and economic influences from both Mexico and the United States. In ways, they may also be considered bicultural.

Regardless of the population growth in Comcáac villages and the persistence of native language use, Seri elders express concern that the cultural knowledge they inherited from their ancestors is endangered, particularly in the domain of natural resources. The degree of retention or loss of such traditional ecological knowledge can be evaluated (Berkes 1999; Zent 1999); for the moment, however, two more fundamental issues concern me: What is truly lost when a native language and its traditional ecological knowledge are lost? And why should a cultural community work to avoid language death and strive to retain such knowledge (Crystal 2000)?

To answer these questions, using the Comcáac as our example, we must first explore how traditional ecological knowledge is transmitted and retained within the community (Nabhan 1998a). We can then consider how this knowledge has had nutritional, medical, spiritual, aesthetic, and social value over the course of Comcáac history, and the ways in which it is still valuable to the contemporary Comcáac. In addition, we can reflect upon how this traditional ecological knowledge might benefit the management of the native flora and fauna of the region and, if appropriately integrated into conservation plans, aid in keeping certain endemic species from extinction. We might also consider the potential benefits of Comcáac knowledge of the medicinal value of endemic plants and animals to other societies, should the Comcáac choose to share such knowledge with others, freely or for a fee. Finally, we might consider assigning additional value to Comcáac knowledge to the degree that it is unique—not "replicated" within other indigenous cultures or in the information warehouses and databases of the Western scientific community.

In the next few chapters I will explore the nature of Comcáac ecological knowledge, particularly as it relates to reptiles, the group of animals with the highest levels of endemism in the Sea of Cortés region. In doing so, I strive for a tangible sense of the tone and detail of how Seri individuals speak about particular reptiles. To do any less would be to perpetuate mere abstract generalities about Native Americans and Nature.

mapping the comcáac sense of place

Seri Homelands and Reptilian Habitats

In geological history, as with that of the people, this is a place of rising and collapsing worlds. . . . The land itself bears witness to the way elements trade places; it is limestone that floated up from the sea, containing within it the delicate, complex forms of small animals from earlier times; snails, plants, creatures that were alive beneath water are still visible beneath the feet. To walk on this earth is to walk on a living past, on the open pages of history and geology.

LINDA HOGAN, *Dwellings* (1995)

I should have been concentrating on helping the two Seri boatmen, but
I felt as though I had frogs in my stomach. Alfredo needed José Ramón
and me to help him launch his panga, a fiberglass dory that is the boat
of choice on the stormy waters of the Sea of Cortés. We had loaded it up
on the beach and were now pushing off for a four-day boat trip to Isla
San Esteban, casting off before the end of the rainy season. The Canal
del Infiernillo was frothing with white-caps. Seeing the rough water, I
thought of the group of Arizona-Sonora Desert Museum biologists who
became stranded on San Esteban a few years ago, losing one of their boats
as its hull was battered against the rocks by a storm. I remembered what
Amalia Astorga said to me when she learned a month ago that I would be
going to the island she called *Coftécöl,* "Giant Chuckwalla": "Don't go out
there unless you are with someone who knows the song to placate Coimaj
Caacoj. He's the giant serpent who lives underwater between Islas Tiburón
and San Esteban. By writhing along on the ocean bottom there, he churns
up the water between the two islands much of the time. If you try to cross
his place between the islands without paying him respect, he'll smash your
boat to bits."

Remembering Amalia's admonition, I asked Alfredo one last thing before we left the mainland: "Is traditional Seri life insurance included as part of hiring you as our guide?"

"¿Seguros?" he repeated in Spanish, confused.

"You know, pues," I said, realizing that I was being too cryptic. "Do you know the song to sing to make peace with Coimaj Caacoj?"

Alfredo and José Ramón looked at each another, then burst out laughing. "I believe I know all the songs we'll need for this trip," said Alfredo, "not just for where that Coimaj Caacoj lives, but for other treacherous places as well."

He jerked the cord to the outboard motor, cranking the engine up to a full-tilt roar. José Ramón gave the panga a last push off the gravelly beach and lifted himself gracefully into the bow. Then we were off, crossing the Canal del Infiernillo as Heerman's Gulls, Eared Grebes, and Brown Pelicans took flight before us or rapidly paddled out of our trajectory toward Isla Tiburón. We skimmed across the straits without hitting much choppy water, for the cross-tides were not yet pulling too strongly. Alfredo effortlessly sped along without a single song crossing his lips. Not all the water we would encounter over the next few days would leave him so silent.

A day later, we pulled away from the southern shores of Isla Tiburón and headed into the pitch black water on our way to San Esteban. Alfredo fixed his gaze on the tidal patterns and wave patches between us and the island to the southwest. He did not steer the boat straight toward San Esteban but stayed just to the east of the most direct passage. I soon learned why.

Alfredo was in his late fifties and had harvested fish and turtles from the waters between the islands for decades. He had always maneuvered these waters in boats equipped with less than optimal equipment—old engines, cracked or bent props, leaking hulls. And yet, by following the routes he had been taught and remembering certain sacred petitions, he had been spared the difficulties to which other, more reckless, pangueros had succumbed. He was now instructing José Ramón in how to read these waters and how to choose a safe route. Long haired, lean, and attentive, José Ramón sat next to Alfredo and silently watched his every move.

Halfway across, I realized that Alfredo was humming a song, barely audible over the outboard engine roar. I caught his eye, and he smiled at me. Just then, a splash of water came over the prow: we had hit a huge patch of rough water, a wave field.

"This is where you must sing to the water serpent Coimaj Caacoj on the roughest days, or else he won't let you make it across," he yelled up to me. Another wave hit, and I turned and looked back at Alfredo and José Ramón; I now noticed a tightness in their shoulders and throats that had not been there earlier.

We entered a series of swells, and water splashed over the bow whenever Alfredo could not avoid being broadsided by a sudden wave. I clenched my fists around the ropes holding the gear down and felt the frogs in my stomach climb toward my throat. Another moment passed as we made it across the most chaotic currents. Considering the choppiness of the surface, we had shipped very little water. Soon, through Alfredo's skill at accelerating to get us onto a wave crest, then gently slipping down the other side, we were out of the serpent's reach, speeding toward the shores of Coftécöl. There, the giant chuckwallas that gave the island its name were visible all the way down to the cobbly beach.

Coimaj Caacoj—the Giant Sea Serpent—had let us pass.

SOMETIME after the world was formed by *Hant Caai*, "Earthmaker," his collaborator, Turtle, dived down to the bottom of the sea and brought up a fistful of mud in the clenched claws of his flipper; with this, he would help Hant Caai make solid ground. But the sea was so deep that it took Turtle months to return to the surface. During that time, some of the mud dribbled out of his claws, so there was not much left stuck to his flipper when Turtle finally surfaced. Nevertheless, Hant Caai knew what to do with it, and the land began to grow and grow (Felger and Moser 1985; Espinosa-Reyna 1997).

"During a time when no land existed, many marine animals tried to reach the bottom of the sea, but had no luck. The only one who finally did so was a giant male Green Sea Turtle, who journeyed three months in order to reach the seafloor. There, he took a big clawful of sand and started to carry it up toward the surface. When at last he arrived, he had only a smidgen left within his grasp, but it was just enough for him to form the mainlands, as well as Isla Tiburón, [as places for] the Comcáac themselves."

FRANCISCO "CHAPO" BARNET, *explaining elements of the Comcáac origin story to University of Hermosillo students (as reported by Zuñiga 1998)*

The trouble was, the mud was still wet when Turtle arrived with it, so he needed the help of *Hant Quizin*, "He Who Toasts the Land." Hant Quizin then enlisted Daddy Longlegs, called *Hant Cmaa Tpaxi Iti Hacáatax*, "One Forced to Venture onto Freshly Formed Ground," to ensure that his work had made the mud firm enough for Hant Caai to walk upon. It was after all this that Hant Caai dreamed up another character, *Hant Iha Quimx*, "Earthspeaker," who is literally called "He Who Tells What There Is on Land."

Hant Iha Quimx, the Comcáac say, was created so that the people could learn all the names of places, plants, and animals. Hant Iha Quimx would repeat the names, for there were so many, it is doubtful that any one Seri individual could ever have kept all of them in his or her head. The Seri dictionary already includes more than 600 place-names (Marlett and Moser 1995), over 290 plant names, and at least 350 animal names in Cmique Iitom (Moser, Moser, and Marlett in prep.). Collectively, these names help create a memorable map for the menagerie of the Comcáac homeland, a territory inhabited by many kinds of plants and animals, including reptiles.

Reptiles wriggle and ramble all over the collective Comcáac map (see Maps 1 and 2). Some place-names remind travelers of areas where venomous or edible reptiles are unusually abundant, while others describe geomorphic features that look like mythic reptiles looming on the horizon. One refers to the Black

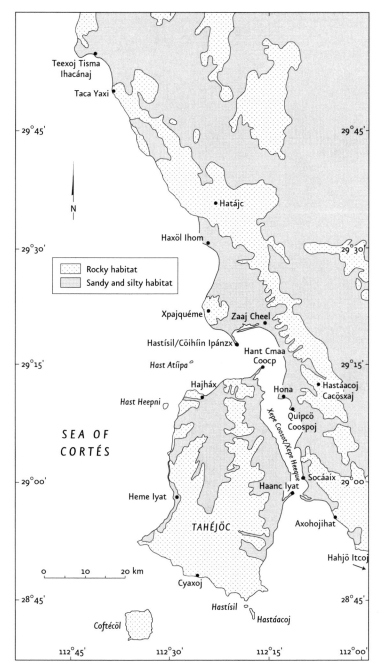

Map 1. Contemporary Comcáac homelands in Sonora, with key place-names in Cmique Iitom, the language still spoken by more than five hundred Seri individuals. Coarse stippling indicates rocky upland habitats used by mountain-dwelling, crevice-hibernating reptiles; finer stippling indicates sandy, silty, and gravelly habitats used by reptiles that excavate burrows in alluvium and aeolian dunes.

ARIZONA

U.S.A.
MEXICO

⊕ Historically visited place

0 ———————— 50 km

Caailipoláacoj
(Laguna Salada)

Hasoj Cheel
(Colorado River Delta)

⊕ Puerto Peñasco

⊕ Caborca

⊕ Pitiquito

SONORA

SEA OF
CORTÉS

Hast Pizal •
(Puertecitos)

⊕ Puerto Lobos

⊕ Cieneguilla

Hast Cheel
(Sierra de Calamajue)

Xazl Iimt (north end)/
Tjamojíil Yacái (south end)
(Ángel de la Guarda)

⊕ Enim (Bacuachi)

Haxöl Ihom
(Desemboque del Sur)

San Miguel
de Horcasitas ⊕   ⊕ Populo

Hast Heepni
(Caiman Rock)

Xepe Coosot/Xepe Heeque
(Canal del Infiernillo)

⊕ Los Angeles

Tahéjöc
(Tiburón) •

Socáaix (Punta Chueca)

Hezitmísoj/Hax Ipac
(Hermosillo/Pitic)

Bahía Kino

PACIFIC
OCEAN

HANT IHIM (BAJA CALIFORNIA)

Coof Coopol Iti Ihom
(San Lorenzo Norte y Sur)

Sosni
(Alcatraz)

Hastáacoj (Dátil /Turner)

Hastísil (Cholludo)

° Iicj Icóoz
(San Pedro Mártir)

Mission Santa
Gertrudis ⊕

Coftécöl
(San Esteban)

Hasoj Iyat
(Guaymas)

BAJA CALIFORNIA NORTE
BAJA CALIFORNIA SUR

Hast Heepni Iti Ihom
(San Pedro Nolasco)

⊕ Belém

Map 2. Greater Comcáac territory, including historically visited places for which oral histories persist among elderly Seri individuals.

Chuckwalla: *Coof Coopol Iti Ihom*, "Where the Black Chuckwalla Is," Isla San Lorenzo. Another refers to a coralsnake: *Coftjéecoj*, "Giant Coralsnake," a hill on Isla Tiburón. Still another, *Cocázni Coopol Iime*, "Black Rattlesnake Where-It-Dwells," refers to a sinuous black volcanic ridge that looks like the giant black rattlesnake that is said to live there; it is just south and east of where Arroyo San Ignacio crosses the road between Puerto Libertad and Rancho Costa Rica, the ranch where W. J. McGee first attempted to describe Comcáac encampments (Fig. 5). And one refers to a rock's similarity in appearance to the profile of a crocodile: *Hast Heepni*, "Caiman Rock," or Islote La Reina, a tidally isolated rock west of Isla Tiburón and a dangerous hazard for boats unaware of its presence. The place-name *Xasáacoj Cacöla* may refer either to the presence of Boa Constrictors or to a sinuous cactus that shares the snake's name and much of its range in coastal Sonora.

Some camps derive their names from the availability of marine reptiles nearby: *Moosni Catxo* "Abundant Sea Turtles," north of Puerto Libertad, and *Moosnipol Quipcö*, "Thick Sea Turtle," on the northeast shore of Isla Tiburón. Isla San Pedro Nolasco near Guaymas, the likely location for the first Comcáac encounters with spiny-tailed iguanas, is called *Hast Heepni Iti Ihom*, "Rock Where the Spiny-tailed Iguana Lives." Still other place-names currently used by the bilingual Comcáac community refer to reptiles by their Spanish names—*Campo Víboras*, "Snake Camp"—though the name in Cmique Iitom makes no such allusion.

Some reptile-associated names for landmarks within historic Comcáac territory may now be used so infrequently that younger Seri have never either heard them or visited there themselves. Such places may be located more than 250 kilometers away from where the people currently live. For example, elderly Seri still recall their names for Puertecitos, Baja California, *Hast Pizal*, "mountain/hill + (untrans.)," and for the Colorado River, *Hasoj Cheel*, "Red River," even though decades have passed since the northernmost Seri group—the *Xica Hai Iic Coii*, "Those Residing toward the True Wind"—last established campsites near the delta of the Colorado.

Although contemporary Seri have little firsthand knowledge of these and other far-off places, as early as 1684 Captain Lorenzo de Bohorques documented a

Figure 5. Ethnologist W. J. McGee initiated modern cultural studies in the central Gulf coast of Sonora at Rancho Costa Rica, where his colleague William Dinwiddie took many photos of a temporary encampment of Seri individuals far inland from extant Comcáac communities of that era. (Photo William Dinwiddie, 1894, courtesy of the Smithsonian Institution)

group of Seris Tepoques—so called for living near Cabo Tepoca, on the Sonoran coast northeast of Isla Tiburón—active at Baquazhi, a well-watered camp inland in the Rio Bacoachi drainage. The "Tepoques" also spent considerable time at Cieneguilla, a wetland in the southern reaches of the Altar Valley, which elders remember but few young Seri visit today. Although a letter written by Bohorques is all that we have to establish the Comcáac presence at Cieneguilla prior to the discovery of gold there in the 1770s (Sheridan 1999), after the gold rush they are known to have frequented Cieneguilla to trade and to intermarry with Hia c-eḍ O'odham, "Sand Papago," families.

Perhaps more remarkable than the persistence of names for distant places is the fact that the Comcáac community is aware of what happens in places hundreds of kilometers from their homes. For instance, Seri fishermen use sightings of enormous dust storms kicked up on the huge playa surrounding Laguna Salada, Baja California Norte, as a prognosis for rough weather caused by Pacific storm fronts. Even though they cannot see the playa itself across the Sea of Cortés, they recognize that the muddied atmosphere northeast of them can only be caused by violent winds scouring the dry silty flats more than 200 kilometers away from their routine fish camps. Seri fisherman Alfredo López remembers the name for this playa as *Caailipoláacoj*, "Great Black Dry Lake."

"You mean the farthest places we know as a people? We know San Felipe in Baja California, and even farther north, up through Mexicali and a large playa, a dry lakebed up there we call Caailipoláacoj. We could see—when a big twister was created by strong winds coming in from the Pacific—the dust [from the playa] rise high in the sky. That was a sight we would use to predict the weather: what happened over there would affect us. Tecate, up in the mountains, we even know our ancestors have gone. They knew those plants and told us about trees there that are not here on the coast or on the islands.

"Then, over that way [northeastward, in the Sonoran interior], we know Cieneguilla, Santana, and Magdalena, where we know particular old trees and cacti that mark where our ancestors have gone. They knew and told us about certain kinds of trees and cacti there that do not grow here on the coast or on the islands. Then to the south, we know and name all the fishing camps down to Guaymas, which we call *Hasoj Iyat*, 'Point at the River.' Do you know that big mountain between Bahía de Cholla, Algodones, and Las Palmas? Yes, the one the Yaquis call *Teta Cahui*, 'Breast Mountain'? We call that *Hast Quiijam*, 'Hill with a View.'

"Beyond that, between Guaymas and Obregón, there are some cactus that we know as among the original ones from which our people ate the fruits. There grow both the *xaasj*, what you call the Cardón, and the *xaasj enim*, which is thinner, with more arms, like the Pitahaya Dulce [Organpipe Cactus]. Yes, it may be what you call *cardón barbón* or *hecho*, the one with the bristles on the fruit. We even know the cacti that far, at a place we call *Xaasj Enim Cöhaníp*, 'Cardón Stuck with a Knife.'"

ALFREDO LÓPEZ OF PUNTA CHUECA,

when asked at Ensenada del Perro about the limits of traditional Comcáac territory

---

It is worth remembering that within the twentieth century, Comcáac fishermen and sea turtle hunters traveled much of the Sea of Cortés. They have boarded cargo boats to travel with Mexican firewood and guano vendors, yachts to travel with American naturalists, and even gigantic cruise ships carrying European tourists. Contemporary Seri also board airplanes and jumbo jets, flying to Mexico City, Paris, and Rome. (And then there was the claim by the late José Astorga that he boarded a spaceship and flew to the moon after he was abducted by Martians, but that, indeed, is another story [Mellado n.d.].)

Comcáac fishermen know the port towns of Baja California's eastern coastline, and the beaches toward Yaqui territory where a group of Seri-speaking people called *Caail Iti Ctamcö*, "Dry Lakes–dwelling Men," once lived. Perhaps these were the remnants of Comcáac dialect groups that missionaries called the Guaymas and the Upanaguaymas (Sheridan 1999). In the 1680s, Captain Encinas reported that Seri-speaking Guaymas people were living among the Cahitan-speaking Yaqui people at the coastal pueblo of Belém. Through such cross-cultural contact, Yaqui fishermen may have adopted the Seri term *moosni* to refer to sea turtles. Perhaps the Yaqui learned at least some of their marine natural history from the Comcáac.

"In the old days, we could go in *balsas* anywhere we can go today in pangas. They enabled us to cross in any season from the mainland, not only to Isla Tiburón, but also to San Esteban, San Lorenzo, and Baja California. Yes, even Baja California was not too far. To the north, we knew Puerto Peñasco, and we have stories about the Río Colorado. To the south, we knew Guaymas, of course, but we sometimes ventured past there. We have a word in our language for central Mexico—how close to Mexico City I don't know—and stories of adventures in the south."

ERNESTO MOLINA OF PUNTA CHUECA, *when asked how far the Comcáac ranged*

Interestingly, some Comcáac familes historically lived away from the Sea of Cortés, dwelling for brief periods at Jesuit and Franciscan missions established among the Yaqui, the Hia c-eḍ and Tohono O'odham, the Lowland Pima, and the Opata or Eudeve. The Franciscan brand of Catholicism was widely adopted by their Uto-Aztecan neighbors. In 1772 and 1773, Padre Gil valiantly tried to make a Franciscan-style mission life for the Comcáac at Carrizal, not far from Bahía Kino, but to put it bluntly, it didn't take (Sheridan 1999). As late as 1802, the Comcáac were still feared for the havoc they wreaked upon mission life, as Padre José Juan de Arrillaga suggested when he blamed them for raiding Santa Gertrudis in central Baja California, nearly 100 kilometers southwest of present-day Desemboque and Punta Chueca (Sheridan 1999):

Some unknown Indians killed an Indian from the mission of Santa Gertrudis and injured others. After committing this outrage, it is believed that they returned to the nearby islands and that they are from the coast of Sonora. . . . My speculations are that they could be Indians from Tiburón Island, situated on the coast of Sonora. They will continue to make hostile incursions by way of their nearby islands as they please or when the weather is moderate, and as they gain knowledge of the water holes which various islands afford between Tiburón and California.

It may well be that the Comcáac did not need to "gain" knowledge of water holes found on the islands between the Sonoran coast and the coast of Baja California, for they had been traveling between the islands for centuries and their oral history accurately recorded where such water holes—scarce or even hidden resources necessary for ultimate survival—occurred. Comcáac place-names often encapsulate the essential color, shape, or size of the land-form immediately above a particular water hole, anchoring it metaphorically in a traveler's mind. Other place-names tell, in a nutshell, of cultural events that occurred in a particular location. These telegrammatic names speak to longer stories, recalling, for instance, a person who became lost and died before reaching that place or a historic struggle that was shaped by that setting. The stories and songs associated with place-names may also have ecological, ethical, or spiritual messages imbedded in them, guiding travelers along "the right path" in a larger sense.

"Our homeground? Well, some outsiders—those that have a certain kind of schooling, I don't know—they say that we came from someplace else. I have heard them say that the Seri, long ago, came from Alaska, but that's not right. We emerged from Isla Tiburón, and for as long as we can remember, we've lived here on the coast, on Tiburón and on San Esteban. We weren't exactly all the same—there were the Tastioteños, the Salineros, and those who came from the mountains on the islands. But we could all understand one another—those of us now called the Seri."

RAMÓN LÓPEZ FLORES OF PUNTA CHUECA,
when asked where the Comcáac originally came from

However expansive this collective cultural geography may be, each Seri individual also responds to the world with a personalized sense of place.

Whether by words or maps, most members of the Comcáac community can define and describe how they personally fit into the collective geography. In the early 1970s, as the Mexican government attempted to legally delineate the Seri *ejido*, or collective corporate territory, two friends of mine, Jim "Santiago Loco" Hills and Hank Hine, asked twenty-four Seri individuals, in Spanish: "Hágame un mapa o dibujo de su terreno, por favor"—"Please make me a map or drawing of your land." They avoided using geographically confining terms such as *lugar* and *hogar*, meaning "place" or "home," which might have encouraged the Seri participants to draw their *solar*, "houseyard," or their *vivienda*, "housing." They also avoided using more global terms such as *mundo*, "world," *tierra*, "earth," or *territorio*, "legally defined territory." Whenever a Seri artist or mapmaker, crayon in hand, asked Santiago Loco or Hank what to put on the blank white paper before them, their response would be "Lo que quiera": "Whatever you want."

"*Ihízitim*—that's *terreno*, homeground. For me, it's Punta Mala on Isla Tiburón; that's where my grandfather lived, where my father, *mi jefe*, was born. It's based around a camp, one that each family has, one that belongs to all of the descendants. It's only for them—those who live there are the only ones who can use the resources nearby the camp. All the descendants have rights there."

JESÚS ROJO, *when asked in Punta Chueca where his homeland is*

As Hine and Hills (2000) commented, there is one feature underlying three-quarters of the maps made for them: "Most maps depict the place of the mapmaker's birth rather than their current place of residence. Of the twenty-four Comcáac mapmakers drawing their land, nineteen depicted an area where they do not reside. Eighteen of these individuals mapped the area of their birth. Ten individuals of this group drew a mark on the map to indicate the exact place where they were born."

We know now that the Comcáac name the exact spot of ground where a child comes out of his or her mother into light, and where his or her iixöni, "placenta," is buried. It is called *Hant Hapx Ihíip*: "Birthing Place."

"See [pointing to a wooden rod stuck into an arm of a columnar cactus, Pitahaya agria]? This was inserted here when a baby was born, and the cactus has grown from the rod upward since then. Many of us have a plant like this, where our parents marked our birth the very day we were born. . . . The very day a child was born here, this cactus was transplanted right where the child's placenta was buried. I still know where my saguaro cactus is at Campo Hona; it was one they carried there for me."

AMALIA ASTORGA, *in the abandoned village of Tecomate on Isla Tiburón, noticing a large Stenocereus gummosus cactus plant outside its typical habitat*

As Hine and Hills (2000) suggest, many of the Comcáac artists they worked with took their question to mean something like "Draw me the shape of your ihízitim." Ihízitim, they learned, is a term shrouded in layers of cultural significance. In the Seri dictionary, Moser, Moser, and Marlett (in prep.) translate it as "terreno de una familia extensa de los antepasados"; in English we might say "land that your extended family associates with your ancestors." It is the place where you are literally connected to your land and your extended family, because your placenta is ritually buried there, by your kinfolk, at the time of your birth. It is where you were raised by a group of interrelated nuclear families who all share the same watering holes and gathering sites (Hills 1973). Though obviously more complex culturally, it is akin to the "home range" concept applied by zoologists and sociobiologists to the area encompassing the local movements of a breeding population of a wild forager.

Each ihízitim is known by a descriptive name taken from its most prominent geomorphic feature—at least, that is what José Juan Moreno told anthropologist William Neill Smith in 1951. Curiously, according to José Juan, these features are used not only to identify one's "homeland" but also to name the male dogs associated with one's extended family group. More recently, I have heard José Juan lecture young Seri men on this same concept. The ihízitim, he clarified, could be named for particular mountains, dry lakes, or water sources, but not for campsites per se.

One historic ihízitim, Coftécöl Iifa, or "Peninsula of the Giant Chuckwallas," was located on the southwestern portion of Isla San Esteban, which, though uninhabited today, is still occasionally visited by Comcáac fishermen. Their brief periods of occupancy are not, however, confused with "title" to the land; that is reserved for a few older men who claim to be descendants of the original inhabitants of San Esteban. As archeologist Tom Bowen (2000) has recorded, those people who historically lived on the island are usually referred to as Coftécöl Comcáac, "People of the Giant Chuckwalla's [Place]." Usually, though, they are known by the descriptive term Xica Hast Ano Coii, "Those [Beings] Who Live among the Mountains," the tag for the dialect group of Comcáac who moved between the rugged southern portion of Isla Tiburón and Isla San Esteban. Some Seri say that they were closely related to "Las Cachanillas," the indigenous folk who inhabited the northeastern coast of Baja California for centuries.

For the Xica Hast Ano Coii who lived within Isla San Esteban's 36 square kilometers, Bowen (2000) has documented four ihízitim. One group centered its activities on Coftécöl Iifa, an area covering only one-eighth of the island's land mass (Villalpando 1989). Within these 4.5 square kilometers, Bowen and archeologist Elisa Villalpando locate the only permanent water hole on the island, Haxáacoj, as well as a hidden cave for vision quests and a handful of other cultural features. Despite its small land area, Coftécöl Iifa has an extensive coastline where abundant intertidal marine life is accessible. It has long supported giant chuckwalla and agave populations dense enough to amply feed an extended family and their dogs.

As Hine and Hills (2000) show, most of the ihízitim drawn by Seri individuals now living on the Sonoran mainland are much bigger than the one located on Coftécöl Iifa, ranging in size from 14 square kilometers to over 250 square kilometers. Hills (1973) demonstrated that the various ecological zones within this hyperarid region vary greatly in their productivity and carrying capacity, both of which rise dramatically as one reaches the coast. He estimated that the amount of land required to support an ihízitim of fifty or more individuals ranged anywhere from 20 to 250 square kilometers, correlating directly with location from arid inland to fog-blanketed coast.

It is probable that certain named places at the edges of Comcáac territory were visited infrequently but still used as landmarks—on the horizon—daily. When Seri individuals map their ihízitim, they demonstrate astounding accuracy in the placement of particular coastal features: islands, sandy spits, dunes, bays, beaches, and coves (Hills and Hine 2000; Nabhan field notes). They sometimes sketch them as if seen from above, looking down at the coast from a nearby mountain ridge, though many are adept at drawing the profiles of mountains, gaps, riverbeds, and canyons from a more horizontal perspective as well. They clearly focus on "landmarks" that can be seen from a distance, ones useful in guiding a boatman on a foggy morning or a hunter on his way home through a dust storm. For instance, from many kilometers west of Isla Tiburón one can see a cliff face unlike any others, for it has a striking chunk of chalky rock just above high tide level. It is called Hast Cooxp Cöquimix Iti Iyat, "Rocky Point with a Lot of White at Its Base," and indicates how far north or south along Tiburón's western shores a seafarer might be.

Of any landscape feature, water holes are perhaps the most frequently named, if only because of their critical importance to survival. There are at least forty named springs, seeps, and water holes on Isla Tiburón, and dozens more on the mainland. One of the most reliable of the water holes on Isla Tiburón is high on a mountainous ridge called Hast Coaaca, "peak + (untrans.)." Yet few Seri individuals visit it today, not because of its remoteness, but for a deeper reason. This water hole is protected by the spring-dwelling serpent Coimaj Caacoj, the terrestrial counterpart of the giant sea serpent that waits with its head near San Esteban and its black tail coiled below Sipoj Iime Quih Iyat, "Osprey Nest Point," a camp on the south side of Tiburón. Some Seri also call the giant serpent up on Hast Coaaca a *corúa*, a hispanicized Nahuatl term; *corúas* are said to protect springs throughout the Americas, from those on the Hopi mesas of the Colorado Plateau all the way down to the Sierra Madre in southern Mexico (Griffith 1989).

Adolfo Burgos was succinct when he spoke of what an angry *corúa* can do: "It tries to eat any people or even any mule deer [*venado buro*] who try to come near the *pozo*, the water hole. Its stomach is full of bits of deer hide and patches of fur. A strong wind roars out of its mouth. It is dangerous to get anywhere near Coimaj Caacoj." This may be the same "Big Snake" that Comcáac elder Santo Blanco drew for authors Dane and Mary Coolidge over six decades ago; Dane heard the accompanying story through a chain of translators, so, like many of the stories he wrote down, it may have come out imperfectly:

> The rattlesnakes have a chief named Big Snake. He lives on Isla Tiburón. He is an enormous rattlesnake and inhabits a cave near the top of the big mountain. There used to be a big water-hole in this cave, but the Chief Snake swallowed a buro deer and then drank up all the water in the pond. He is longer than from here to the Big House [one hundred and fifty yards] and he is fatter than this tent [twelve feet]. [Probably a black water-boa.] When he is hungry he goes into the brush, dragging in to him with his tail all the animals in the circle. Sometimes he eats eight to twelve buros at one time. (Coolidge and Coolidge 1971)

The Big Snake described by Santo Blanco must be the one the Comcáac community calls Coimaj Caacoj, but he was not necessarily a rattlesnake or a "chief." Nor was it the only time that Dane Coolidge casually recorded a story about water serpents in Sonora. In his book *Texas Cowboys*, Coolidge (1981) tells of an old Mexican cowboy, Manuel, who made a "great deal about a big black snake down in Sonora that had a golden cross on its head. It lived in a cave on the mountain and the people offered it milk and young chickens and worshiped it like a god."

I recently rode horses with a Sonoran cowboy near Bahía San Carlos who said that only rarely does he see *corúas* out in the *monte* hunting because they spend most of their time waiting alongside seeps and springs in the Sierra El Aguaje. Cultural geographer Ralph Beals (1945), who lived and worked among the Yoemem in the 1930s, reported a similar story associated with certain desert springs in the Sonoran foothills: "The Yaqui and Mayo of Sonora believe horned serpents live in springs in the mountains. These springs never go dry. When the serpents leave the springs they go down the rivers to the sea, causing the floods which are important to the agriculture of both peoples. A young Yaqui informant had his canoe turned over by a water serpent passing beneath him as he crossed the river during the flood of 1928."

As both Beals (1945) and folklorist Jim Griffith (1989) point out, many Uto-Aztecan farming cultures relate *corúa*-like water serpents to the protection of springs critical to crop irrigation, and to floods essential for renewing the soil fertility of their *milpa* maize fields. *Corúa* and *serpiente* have also become terms used by irrigation agriculturists in Sonora to describe everything from the floods themselves to curvilinear water chutes used to connect irrigation canals. As Griffith (1989) concludes, "Whatever *la corúa* is, however it came to be embedded in our regional culture, it is a creature of the New World rather than an importation from the Old."

Clearly, a serpent that protects water holes may be just as essential to the Comcáac traditions of hunting, gathering, and seafaring as it is to Uto-Aztecan farming traditions. Coimaj Caacoj lives high in the mountains, protecting the last water available after droughts have dried up all other reservoirs where Comcáac community members can quench their thirst. If not spared from overexploitation by threats of death from the most frightening creature imaginable, this essential resource, too, might dry up. The serpent also protects Isla San Esteban's sole water hole from those who come to exploit its resources from the mainland or from Isla Tiburón.

Each ihízitim had at least one water hole in a location remote from the habitual encampments, a critical buffer against thirst during drought periods, which in Comcáac country can last anywhere from four to twenty-six months. The Comcáac must not only learn the exact locations of these springs and *tinajas*, but also respect them and keep them from becoming depleted or contaminated.

Sometimes it takes a big snake to keep the need for such protection paramount in the minds of a community.

the shape of reptilian worlds

Island Biogeography and the Herpetofauna of the Sea of Cortés Region

[Herpetologist Ted] Case is briefing us on the geography of the Gulf of California. He's using freshly fried chips as tokens, and Ángel de la Guarda is represented by a large one with an elegant curl. "Here's Esteban," he says. Another island, another chip. "And these are the Lorenzos"—two other islands on which he has also done fieldwork— "and Partida." We stare down at five insular bits of crispy tortilla, awash in a sea of tabletop. The bottles of Pacifico beer could be a fleet of fishing boats. "The problem is," says Case, "all these islands create crazy patterns of tide and wind moving between them. It can get very squirrelly."

. . . Case studies giant lizards on the islands of the Gulf of California. But don't be misled. The real subject, again, is the shape of the world.

DAVID QUAMMEN, *The Song of the Dodo* (1996)

The gulls were shrieking over our heads as a Mexican panga carried us out of the harbor at the fishing village of Kino Viejo. The wind was up and so were the waves—it took us thirty minutes, nearly twice as long as it would normally take to reach Isla Alcatraz. At last we reached the western shore of the small guano-covered islet, just as the scorching sun dropped below the horizon of Isla Tiburón some 50 kilometers away. That cooled us off, but left us little daylight for finding what we were looking for: a giant chuckwalla with five times the body mass as its kin on the Sonoran mainland across the bay. I scurried over a cobbly beach littered with oyster shells, clams, and manta ray carcasses, then scrambled up the slopes, looking for shallow caves and ledges amid the volcanic boulders.

The slope was covered with little plant life other than giant cardóns and bristly cholla. I carefully sidestepped these sharp-spined cacti, crawling on my hands and knees over and around clusters of boulders hoping to find animals in the shallow caves beneath them. The first few had small crevices beneath them, but they contained only scat, no live creatures. I moved on, a little desperate, climbing steeply uphill.

After a certain point, the air began to cool, and a rain of crepuscular pollen started to move like a mist down the slope. The pollen rain brought on a sneeze, which made me shift my scrambling posture. Straightening up, I saw a Great Blue Heron, frightened by my sneeze, take flight from a cardón. Then I noticed something stirring under a ledge of fractured volcanic tuff a few meters to my right. I leaned over, peeked around a cholla, and caught sight of a chuckwalla's hind end. Its tail had recently dragged across the silty bench in front of the ledge, its characteristic imprint easily visible even in the failing light.

I quietly crawled toward the resting chuckwalla, then advanced upslope from it, where I carefully began removing cobbles on either side of the ledge it was hidden under. As I dislodged one cobble, the partially exposed chuckwalla stirred, revealing another, smaller, chuck tucked in safely behind it.

After taking a deep, quiet breath, I made my move. My right hand clasped the bigger chuckwalla, a male, just in front of his back legs. He tried to dig further in under the ledge, but I lifted his lumpy black-and-buff body up into the air before he could lodge himself deeper into the crevice. He flailed his chubby legs for a moment, but then, as I stroked his underbelly, he grew calm and tame.

He was the size of a Sunday newspaper rolled up and thrown onto your porch. That made him far larger than any I had seen on the mainland, and he had many more black blotches than those I recalled from my earlier trip to Isla San Esteban with two Seri men from Punta Chueca. I remembered how Alfredo López had held one of the Piebald Chuckwallas there, looking at its buttonlike femoral pores on the back legs, which he likened to poker chips.

"Yes, this one is a male," he said, laughing and pointing to the swollen, hormone-charged pores lining the crease of its back leg. "He's *muy macho;* see his winnings?" He told a story about chuckwallas being good gamblers and winning their femoral-pore "poker chips" from other inhabitants of San Esteban.

I held my own chuckwalla up in the dusky light and saw that five or six large pores were oozing goo out along the ridges on each back leg. This confirmed that the chuck was indeed a male, but I was not sure what species he was. He looked as though he had gathered traits from several different species, like the melting-pot descendants of immigrants who had passed through Ellis Island off New York.

Whatever his species, this giant was *my* "winnings" for the night. Catching him was no big deal, but spending time in the hunt had helped me better understand a creature whose history is intimately linked to that of the Comcáac. I sensed that the Comcáac had created a distinctive "breed" of chuckwallas, just as much as the Navajo had shaped their own breed of Churro sheep, and Australian aborigines had shaped the course of dingo evolution.

My friends hollered for me to get back to the boat, and I placed the chuckwalla on a flat rock still warm from the day, released it, and watched as it lumbered off into the darkness. Once I and my party headed home, vacating the island, the chuckwalla again became the largest flightless land animal on Isla Alcatraz, the 1,800-gram king of the mountain.

IN MANY LANGUAGES, the word used to describe human households is also used metaphorically to describe the places where reptiles and other animals lay their eggs, rest, mate, or congregate. The Comcáac speak of *iime*, a word now used to refer to houses constructed of concrete blocks, pre-fab cement panels, tar paper, or tin. Historically, *iime* referred to ocotillo frames covered with brush, hides, or turtle carapaces, makeshift shelters that one to three people could crawl into (Fig. 6). As ethnographer W. J. McGee (1971) described them in 1894,

> the walls are thicker than the roofs, which are supplemented in the time
> of occupancy by haunches of venison, remnantal quarters of cattle and
> horses, half-eaten turtle, hides and pelts . . . irregularly clap-boarded
> with turtle-shells and sheets of a local sponge. . . . The Seri habitation
> is not a permanent abode, still less a domicile for weaklings or a shrine

for household lares and penates, not at all a castle of proprietary sanctity,
and least of all a home; it is rather a time-serving lair than a house
in ordinary meaning.

Such shelters were constructed and used whenever the Comcáac needed a temporary refuge from the scorching summer sun, the winter cold, or the frequent
winds that would otherwise whip sand against their skin. The buildings found
in Comcáac villages today appear much more like "houses in ordinary meaning," but they tend to be used as buffers from extreme cold and heat more than
as the spaces central to family activities; most cooking, sleeping, craftsmaking,
and socializing go on "outside" rather than "inside" these shelters.

Iime is a multifaceted term, and its meanings are layered (Moser, Moser, and
Marlett in prep.). Although it has been casually translated as "house" or "housing," its root meaning may be closer to "dwelling," "shelter," "refuge," or even

"nest." Indeed, this term is regularly employed when speaking of birds' nests and their locations, as several place-names demonstrate. *Sipoj Iime Quih Iyat*, "Point Where the Ospreys Built a Nest," for example, is the name of a camp on a jetty along Isla Tiburón's coast, and *Hanaj Iime*, "Raven's Roost," is a camp on Tiburón as well.

Iime can also be used to refer to the nests, shelters, or congregating places of reptiles. For Desert Tortoises, Regal Horned Lizards, Gila Monsters, and various rattlesnakes, chuckwallas, and iguanas, these nesting spots are often small caves or shallow excavations under ledges. There, desert reptiles may rest, tend to their eggs or young, or go into seasonal torpor. Desert Tortoises and Western Diamondback Rattlesnakes may share a shelter, as demonstrated by a shallow burrow that Guadalupe López pointed out to me on Isla Tiburón. These reptiles may also estivate or hibernate with other species in the same shelter (Table 5). Reptile hibernation or estivation is what the Comcáac refer to as *cpoin*, "shut-down time" or "closed-in time."

"In the time of heat, none of the sea turtles need to cover themselves [when they rest] with the fleece-like layer we call *xpanáams* [marine algae]. But in the winter, certain ones will put algae on their carapaces so that they won't suffer from the cold. They will dig themselves into their *iime* until you can see only half their carapace, nothing else of them. We call the places where they do this *iime* or *moosni iti hax yaiij*—the same places some of the Mexican turtlers call *surgideros*."

GUADALUPE LÓPEZ, *Desemboque*

Marine reptiles are recognized to have *iime* as well. Many of the *moosni iime* the Comcáac describe are along undersea troughs, ten to thirty meters deep, that serve Green Sea Turtles as both summer retreats and wintering places. Some *moosni iime* are situated where natural upwellings (*surgideros*) occur. All are rich in the food resources that sea turtles need. The muddy and sandy bottoms of troughs edging eelgrass stands are where the *moosni hant coit*, "sea turtles touch-

TABLE 5

Terrestrial Reptiles Reported by Comcáac to Share *Iime* Shelters

Species	English Name	Co-inhabitants
TESTUDINIDAE		
Gopherus agassizii	Desert Tortoise	Sonoran Coachwhip, Gila Monster, Desert Spiny Lizard, Western Diamondback Rattlesnake,[1] packrats
PHRYNOSOMATIDAE		
Sceloporus magister	Desert Spiny Lizard	Desert Tortoise
HELODERMATIDAE		
Heloderma suspectum	Gila Monster	Desert Tortoise, rattlesnakes
COLUBRIDAE		
Trimorphodon biscutatus	Lyresnake	(Lesser Long-nosed) Bat[2]
VIPERIDAE		
Crotalus spp.	rattlesnake species	Gila Monster, Desert Tortoise,[1] other rattlesnakes

[1] On September 11, 1999, on Isla Tiburón at Ensenada del Perro, Guadalupe López showed me and Craig Ivanyi a burrow with recent tortoise draggings under an ironwood tree. When Ivanyi investigated the burrow, he discovered that a Desert Tortoise and a Western Diamondback Rattlesnake were resting within 15 centimeters of one another; they continued to do so the entire time we were present. The shallow burrow looked as though it was used as a temporary refuge from midday heat during the summer; larger crevices in rocky outcrops 100–300 meters away might better serve for longer hibernation.

[2] In Degenhardt, Painter, and Price 1996, Mark Doles and Douglas Burkett reported seeing *T. biscutatus* hibernating in a rock crevice with *Crotalus atrox*, *C. lepidus*, and *C. molossus*.

ing down," go into winter dormancy. They are also the primary places where Green Sea Turtles were hunted.

In the 1970s my former housemate Kim Cliffton made many trips with Comcáac turtle hunters across the Canal del Infiernillo. He was amazed to learn that each iime was called by a specific traditional name and that the hunters in their pangas could precisely locate each of these places by triangulating from landmarks onshore (Cliffton, Cornejo, and Felger 1982). As Felger and Moser (1987) put it, "There were numerous places where the Seri believed sea turtles lived, each of which was named as a different turtle iime, 'home,' or resting place. Even at night hunters were able to align themselves [to find a particular iime] with specific mountain peaks." In the 1990s I asked Comcáac hunters if they continued to visit the *moosni iime* within their reach. And indeed, they still maintained knowledge of at least twenty-six such localities between Bahía Kino and Desemboque (Table 6). Guadalupe López and other seafaring Seri, for example, were still able to pinpoint where they could find *Moosníctoj Iime*, "Reddish Turtle's Resting Place," near Punta Santa Rosa (providing the shortest reach between Isla Tiburón and the mainland), even though they were no longer harvesting turtles commercially. By far the greatest density of *moosni iime* is in the central stretch of the Canal del Infiernillo.

It is a sad fact that during the late 1970s and 1980s, one after another of these sea turtle retreats was depopulated. Whereas Comcáac turtle hunters had traditionally harvested some sea turtles within the cluster of *iime* in the Canal del Infiernillo, they did not deplete them. Then in 1975, their non-Indian neighbors began to find dormant sea turtles both in the Canal and elsewhere (Felger, Cliffton, and Regal 1978). They relentlessly overexploited these vulnerable refuges throughout the midriff of the Sea of Cortés until the early 1990s, when the Mexican government's ban on selling sea turtles and their products began to be more widely enforced. This overexploitation did not make the sea turtles "homeless" so much as it made many of the traditional *iime* resting grounds "turtle-less." Today, Guadalupe López claims that the turtles can no longer stay in these winter resting grounds because motorized boats regularly disturb their slumber.

TABLE 6
Green Sea Turtle Gathering Grounds Recognized by the Comcáac

From Bahía Kino northwest to south end of Isla Tiburón (at Ensenada del Perro)

Hast Coopx
Hast Quicös
Soosni Iime/Quipcö Iti Icatóoyaj

From south end of Isla Tiburón to Punta Chueca

Xnit Iime/Zaaj
Zaalca Yax Iime
Moosníctoj Iime
Cmaam Quiscáma Quih Iti Hascáma Ihíip
Moosni Imoquéee
Moosni Cahíxöt/Pnaacoj Cacösxaj Hapx Ihíip
Hant Icahéme
Socáaix Iime

From Punta Chueca north to Punta Arena

Xtaasi Iime
Xepe Imac Iime
Coníic Iime
Sacpátix Iime
Ziipxöl Iifa Xatj
Yacáptax Iime
Iime Heeque
Hatám Caacöl
Tamíha
Pasítaj Iime/Xtasíit

From Punta Arena north to Desemboque del Sur

Iime Cooxp
Sana Iime
Xatj Colx Quiij
Xojtís Iime
Hastéexp Iifnij Hapx Ihíyat

Comcáac elders tell stories and even sing about certain hunting places in the Canal del Infiernillo. One song commemorates a rock off the shore of Campo Palo Fierro, where underwater caves, drop-offs, and upwellings attract congregations of sea turtles who come "to play" there. Comcáac oral histories distinguish these upwellings from *iime* used by winter-dormant or resident individuals and from those used primarily by the *cooyam*, the developmental stage of the Green Sea Turtle in which it "migrates from afar" or "travels from the high seas."

Another kind of sea turtle "nest" is unfortunately seen today by fewer and fewer Seri individuals. Only a handful of contemporary Comcáac have ever encountered the egg-laying of sea turtles on beaches or dunes near their homes. Although Guadalupe López is sure that he saw a female Leatherback lay eggs once when he was young, and there are confirmed reports from the late 1970s of Leatherback nestings in Baja California, most of the elderly Comcáac beachcombers who have stumbled upon eggs being laid during the day on Isla Tiburón and Bahía Kino have probably seen olive-colored Pacific Ridleys. Each year, a few individuals of this species are known to bury their eggs away from the massive *arribadas* on the prime nesting beaches more than a thousand kilometers to the south of Comcáac territory, venturing to Baja California and as far north as Puerto Peñasco on the Sonoran mainland (Fritts, Stinson, and Marquez 1982; Navarro 1997; G. Montgomery, pers. comm.).

"Once, when I was young, I saw sea turtles—what kind I'm not sure—come up on the sandy beach at Bahía Kino. It was their *arribada*. During the night, I saw them come up onto the beach, and with their flippers, the females dug in the sand until their carapaces were halfway buried, and there they laid their eggs. And later, when the eggs were hatched—eggs like those of hens—there came wriggling out into the open tiny baby turtles, no more than four or five inches long. And then [he waved his hand off toward the high seas] they left with the tides."

RAMÓN PERALES, *retired sea turtle hunter, Punta Chueca*

To say that there are Seri individuals intimately familiar with the locations of reptile nest sites is to understate the matter. Between 1978 and 1982 biologists S. Reyes-Osorio and R. B. Bury conducted Desert Tortoise surveys of Isla Tiburón, working together with Seri field assistants, and in so doing they claimed to have recorded one of the highest densities of Desert Tortoises ever found. One of their Seri fieldhands, Alfonso Méndez, later told me that he worked as their "finder" of Desert Tortoise congregating places. He recalled finding several hibernating groups—one numbering no less than seventy-five individuals holed up in a single cave—that the biologists told him were much larger than any they had ever found on their own (Reyes-Osorio 1979).

"A Desert Tortoise was walking on a hill when he came to a halt as he was beginning to eat something. Then, from afar, there came hunters arriving to kill him. Now there is danger—he has to hide. There will be danger if the hunters get to see him—for this, he has to hide."

ALFREDO LÓPEZ, *explaining a Comcáac song*

Whether this Seri hunter simply had a knack for tracking down hibernating colonies, or whether he had good recall of places where other Seri had formerly found them, I do not know. However, in spring 2001, Méndez joined me and my students on Isla Tiburón and relocated several tortoise caves he had seen previously. Historically, both Comcáac women and men would seek out these iime shelters with the aid of dogs, which could smell the hibernating Desert Tortoises long before humans could see them. (The exact location of some iime for chuckwallas, tortoises, and rattlesnakes could also be deduced by tracing the tracks and tail draggings left on sand- or silt-covered benches back into the rocks where these animals were sequestered.) Since any given area has only a few suitable shelters, these iime were likely used by generations of tortoises over many centuries. The Comcáac community has consequently kept in its collective memory a living record of where particularly large congregations of Desert Tortoises can be encountered during extended periods of cold or drought.

The value of teaching younger generations which "Desert Tortoise caves" also harbored a Western Diamondback Rattlesnake or a Gila Monster should be obvious as well. Although these venomous reptiles might not bother their cohabitants, they could bite or frighten anyone naively arriving to capture tortoises.

The Comcáac "shape of the world" has been contoured by where these *iime*, both marine and terrestrial, occur. When Comcáac hunters walk into arroyos and up ridges where they have formerly encountered reptiles and other wildlife, these spots—ones where desert tortoises were found in caves by their grand-parents' dogs or where whipsnakes climbed to the top of enormous stands of mesquite—are already marked in their minds. As they cross the channel be-tween the mainland and Isla Tiburón, they may come upon an eddy or a rip current hardly noticeable to the likes of you or me, but which alerts them to a particular *moosni iime*, perhaps the one where their brother harpooned his first sea turtle.

THE ACCUMULATION of experiences and oral histories about the *iime* creates a cognitive map of where each kind of animal characteristically nests, rests, for-ages, and hibernates. The deserts and waters are not "open spaces" as much as they are culturally shaped *geographics*, worlds that are "marked" in people's mem-ories through stories, songs, and sayings. The Comcáac themselves have helped shape the geographic distributions of certain reptiles by intentionally or unin-tentionally dispersing them to islands, by hunting them, and by providing new shelters or food resources for them.

To gain a sense of Comcáac influences on reptilian diversity, we must first make some assumptions about what species occurred on the coast of Sonora and adjacent islands prior to occupancy by Comcáac communities. We can then record any introductions or immigrations, as well as local extirpations or global extinctions. With such interpretations, we can surmise whether Comcáac cul-tural activities have in fact restricted or broadened the ranges of particular species.

This exercise may at first seem simple, but it is complicated by our imper-fect knowledge of the prehistoric, historic, and current distributions of the region's herpetofauna. It is nevertheless an important exercise to attempt, in part because it may demonstrate to quantitative biogeographers how seldom they have factored in the influences of indigenous peoples on patterns of faunal diversity and abundance. As you read the rest of this chapter, keep in mind the following passage from a major review article in the *American Scientist* by island biogeography theorists Ted Case and Martin Cody (1987):

> In practice it is next to impossible to find a set of islands without environmental change or evidence of the severe impact of man's activities. The islands of the Sea of Cortés probably come closer than any to fulfilling the requisite conditions. There was never any permanent aboriginal population except on Tiburón, and the presence of modern man is restricted to a few small settlements in three of the larger islands. Relatively few islands have any plants or animals introduced from elsewhere.

Over the last century of scientific exploration of the Sea of Cortés region, biologists have recorded fifty-one terrestrial reptile species on the central Gulf coast of the Sonoran mainland and on the adjacent islands of Tiburón, San Esteban, Cholludo, Dátil, and Alcatraz (Table 7). These fifty-one species inhabit much of what is known as the central Gulf coast biotic community, a subdivision of the Sonoran Desert, with a distribution closely matching the Comcáac territory of the last two centuries. The habitats in this central Gulf coast region, which receive only 100 to 200 millimeters of rainfall in most years, range from ancient cactus forests, with three to five species of columnar cacti and ironwood dominating, to scrublands, where elephant trees, limberbushes, saltbushes, and brittlebushes form mixed stands without a lower shrub layer beneath them. Mangroves and riparian gallery forests along watercourses, though now restricted to just a few sites within this hyperarid, nearly frost-free region, formerly played important roles in maintaining a number of wet tropical species at their northern limits.

Of course, the Comcáac historically ranged into other biotic communities:

the plains of Sonora desert scrub, foothills of Sonora thorn scrub, the lower Colorado River Valley, and perhaps even the Vizcaíno region of central Baja California. If we also consider the possibility that they encountered at least fifteen other reptiles endemic to Islas San Lorenzo, San Pedro Mártir, Partida Norte, San Pedro Nolasco, and Ángel de la Guarda, our list of desert reptiles potentially encountered by the Comcáac would extend to well over sixty species (Table 8).

Another six marine reptile species (five sea turtles and one seasnake) have been recorded by biologists in tribally managed waters. In addition, most biologists I have spoken with accept that crocodiles may have occurred at least as far north as Guaymas and the Río Yaqui a century ago, as Richard Felger suggests in a 1998 newsletter essay (Table 9).

Of the fifty land species of reptiles found in contemporary Seri homelands, at least five now dwell in locations where they do not necessarily appear to occur naturally: San Esteban Spiny-tailed Iguanas; Black, Desert, and Piebald Chuckwallas; and one or more leaf-toed geckoes (Map 3). Biogeographers doubt whether either natural long-distance dispersal or vicariance (inundation of land bridges) can reasonably account for some of these unusual distributions.

As a case in point, herpetologist Lee Grismer (1994a) has concluded that the presence of spiny-tailed iguana populations on Islas San Esteban and Cholludo "is probably a result of aboriginal introduction by Seri Indians who once inhabited the island. The coastal distribution of C[tenosaura] hemilopha on mainland Sonora only extends as far north as near Guaymas, approximately 115 km south of the nearest coastal mainland locality opposite Isla San Esteban."

This distributional anomaly is matched by an ecological anomaly: San Esteban is the only island out of more than one hundred in the Sea of Cortés where both spiny-tailed iguanas and gigantic chuckwallas occur together (Fig. 7). Contrary to predictions from ecological niche theory, they successfully coexist there, foraging for largely the same plants in the very same habitat, at the same time of day, over the the same season. Grismer (1999) and his students report DNA evidence which suggests that Isla San Pedro Nolasco, west of Guaymas, is the most probable source of the iguanas on San Esteban.

TABLE 7

Reptiles of Central Gulf Coast and Midriff Islands Nearest Comcáac Villages

Species	Mainland Sonora	Tiburón	San Esteban	Cholludo	Dátil	Alcatraz
TURTLES AND TORTOISES						
TESTUDINIDAE						
Gopherus agassizii	•	•			•?	
KINOSTERNIDAE						
Kinosternon flavescens	•					
K. sonoriense	•					
LIZARDS						
CROTAPHYTIDAE						
Crotaphytus dickersonae		•				
C. nebrius	•					
Gambelia wislizenii	•					
IGUANIDAE						
Ctenosaura conspicuosa			•	•		
Dipsosaurus dorsalis	•					
Sauromalus obesus	•	•				
S. hybrids						•
S. varius		•?	•			
PHRYNOSOMATIDAE						
Callisaurus draconoides	•	•				
Holbrookia maculata	•					
Phrynosoma solare	•	•				
Sceloporus clarkii	•	•				
S. magister	•	•				
Urosaurus graciosus	•					
U. ornatus	•	•				
Uta stansburiana	•	•	•		•	•
EUBLEPHARIDAE						
Coleonyx variegatus	•	•				
GEKKONIDAE						
Phyllodactylus homolepidurus	•?					

Continued on next page

TABLE 7 (continued)

Species	Mainland Sonora	Tiburón	San Esteban	Cholludo	Dátil	Alcatraz
P. xanti		•	•	•	•	•
TEIIDAE						
Cnemidophorus burti	•	•				
C. tigris	•	•	•			
XANTUSIIDAE						
Xantusia vigilis	•					
HELODERMATIDAE						
Heloderma suspectum	•					
SNAKES						
BOIDAE						
Boa constrictor	•					
Charina trivirgata	•	•				
COLUBRIDAE						
Arizona elegans	•					
Chilomeniscus stramineus	•	•				
Chionactis occipitalis	•					
Hypsiglena torquata	•	•	•			
Lampropeltis getula	•		•			
Masticophis bilineatus	•	•				
M. flagellum	•	•			•	
M. slevini			•			
Oxybelis aeneus	•					
Pituophis melanoleucus	•	•				
Rhinocheilus lecontei	•					
Salvadora hexalepis	•	•				
Trimorphodon biscutatus	•	•				
ELAPIDAE						
Micruroides euryxanthus	•	•				
HYDROPHIIDAE						
Pelamis platurus	•	•	•[1]			
VIPERIDAE						
Crotalus atrox	•	•			•	

TABLE 7 (continued)

Species	Mainland Sonora	Tiburón	San Esteban	Cholludo	Dátil	Alcatraz
C. cerastes	•	•				
C. estebanensis			•			
C. molossus	•	•				
C. scutulatus	•	•				
C. tigris	•	•				

CROCODILES

CROCODYLIDAE

| Crocodylus acutus | • | (•)[1] | | | | |

Source: Grismer 1999.

[1]See Table 9 and text.

TABLE 8

Endemic Reptiles of Other Midriff Islands Visited by Comcáac Seafarers

Species	Island
LIZARDS	
CROTAPHYTIDAE	
Crotaphytus insularis	Ángel de la Guarda
IGUANIDAE	
Ctenosaura nolascensis	San Pedro Nolasco
Dipsosaurus dorsalis	Ángel de la Guarda
Sauromalus hispidus	Ángel de la Guarda, San Lorenzo Norte and Sur
S. obesus	Espíritu Santo, Partida Sur
S. varius	Roca Lobos
PHRYNOSOMATIDAE	
Callisaurus draconoides	Ángel de la Guarda, Espíritu Santo, Partida Sur, Patos
Sceloporus clarkii	San Pedro Nolasco
Urosaurus nigricaudus	Espíritu Santo, Partida Sur
Uta nolascensis	San Pedro Nolasco
U. stansburiana	Ángel de la Guarda, Patos, San Lorenzo Norte and Sur

Continued on next page

TABLE 8 (continued)

Species	Island
EUBLEPHARIDAE	
Coleonyx variegatus	Ángel de la Guarda, Partida Sur
GEKKONIDAE	
Phyllodactylus homolepidurus	San Pedro Nolasco
P. partidus	Partida Norte
P. unctus	Espíritu Santo, Partida Sur
P. xanti	Ángel de la Guarda, La Raza, San Lorenzo Norte and Sur
TEIIDAE	
Cnemidophorus bacatus	San Pedro Nolasco
C. espiritensis	Espíritu Santo, Partida Sur
C. martyris	San Pedro Mártir
C. tigris	Ángel de la Guarda, Espíritu Santo, Patos, San Lorenzo Norte and Sur
SNAKES	
BOIDAE	
Charina trivirgata	Ángel de la Guarda
COLUBRIDAE	
Chilomeniscus punctatissimus	Espíritu Santo, Partida Sur
Hypsiglena torquata	Ángel de la Guarda, Espíritu Santo, Partida Sur, San Lorenzo Sur
Lampropeltis getula	San Pedro Mártir
Masticophis barbouri	San Pedro Mártir
Salvadora hexalepis	Espíritu Santo
VIPERIDAE	
Crotalus atrox	San Pedro Mártir
C. enyo	Espíritu Santo, Partida Sur
C. mitchellii	Espíritu Santo, Partida Sur, Salsipuedes
C. ruber	Ángel de la Guarda

Source: Grismer 1999.

As for the San Esteban Spiny-tailed Iguanas on Isla Cholludo, the most plausible explanation for their arrival there is cultural dispersal from Isla San Esteban 20 kilometers away (Grismer 2002), since they are not found even 1 kilometer away on Isla Dátil or on Isla Tiburón. Pioneering herpetologists Chuck Lowe and Ken Norris (1955) first suggested that both chuckwallas and iguanas

TABLE 9

Marine Reptiles Historically Reaching into the Sea of Cortés

Species	English Name	Spanish Name
TURTLES AND TORTOISES		
CHELONIIDAE		
Caretta caretta	Loggerhead Turtle	Caguama cabezona, jabalina
Chelonia mydas	Green Sea Turtle	Caguama carrinegra
Eretmochelys imbricata	Hawksbill	Carey, perico
Lepidochelys olivacea	Pacific (Olive) Ridley	Golfina
DERMOCHELYIDAE		
Dermochelys coriacea	Leatherback	Siete filos
SNAKES		
HYDROPHIIDAE		
Pelamis platurus	Yellow-bellied Seasnake	Alicante del mar, anguilla
CROCODILES		
CROCODYLIDAE		
Crocodylus acutus	River Crocodile	Cocodrilo, caimán

"probably reached these islands by transport by man or by birds. The Seri Indians and some Mexicans eat both of these lizards." Ironically, over the following decade not a single herpetologist spoke directly with the Comcáac to try and verify this hypothesis, and for the most part quantitative biogeographers have continued to ignore its implications altogether. Nevertheless, herpetologist Michael Robinson (1972) proposed that the chuckwallas on Isla Alcatraz were of hybrid origin mediated by cultural dispersal, and gradually evidence has accumulated showing that traits of three species of *Sauromalus* have been found together in one population on Alcatraz. Recently, other herpetologists have suggested that cultural dispersal is the most plausible explanation for the presence of Piebald Chuckwallas on Isla Roca Lobos near Isla Salsipuedes, and of Black Chuckwallas nearly identical to those of Ángel de la Guarda on Islas La Ventana and Smith off the shore of central Baja California (Grismer 1994a; Hollingsworth et al. 1997; Petren and Case 1997).

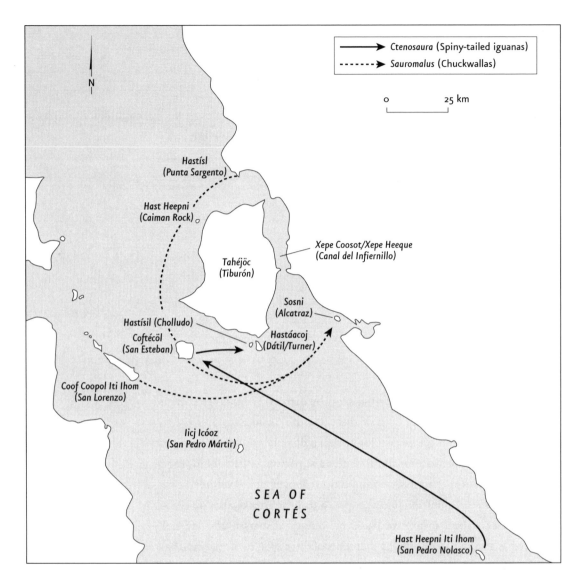

Map 3. The cultural dispersal of iguanids between the midriff islands, as hypothesized from Seri oral histories and recent genetic and morphological analyses.

Figure 7. A chuckwalla carved from palo blanco by José Luis Blanco. (Photo Helga Teiwes, n.d.)

Whether this hypothesis is called "cultural dispersal" or "aboriginal introduction," only three scientific papers from before 1970 considered it a viable factor explaining biogeographic patterns in the Sea of Cortés; they were supported in the 1970s, however, by Michael Robinson's careful work (unpubl.). In the 1980s Ted Case threw his lot in with the others who saw cultural dispersal as a significant factor in chuckwalla biogeography, as did Lee Grismer, Ken Petren, Howard Lawler, and Charlie Silber in the 1990s. In any event, in the past fifty-odd years it has become clear that indigenous seafarers do not need to have permanent settlements on islands to have reason to disperse reptiles to them.

"During the era when commercial totoaba [a large corvina] fishing began, the Seri carried chuckwallas to Alcatraz so that they could be more accessible when there was no other food available. Today, we don't need survival foods as often, so we let them [the chuckwallas] be."

ALFREDO LÓPEZ, *Punta Chueca*

Curiously, a leaf-toed gecko from the Baja California peninsula was apparently dispersed to Islas San Lorenzo, San Esteban, Tiburón, and Alcatraz, perhaps in a way that contradicts the early belief of Soulé and Sloan (1966), two dogmatic island biogeographers who insisted that little mixing took place between peninsular faunas and those of the easterly midriff islands. Lee Grismer (pers. comm.) pointed out that, to the contrary, this gecko clearly exemplifies such "faunal mixing," but he could find no evidence to attribute its anomalous distribution to intentional introductions by aboriginal inhabitants. Desert herpetologist Phil Rosen (pers. comm.), however, has observed that geckoes are the quintessential "camp followers," capable of hitching rides with humans by hiding in their boats or baggage as stowaways. It is well documented, for instance, that Polynesians inadvertently introduced four species in four different genera of geckoes to the Hawaiian islands prehistorically (Austin 1999; McKeown 1978). Literary herpetologist Maurice Richardson (1972) has argued that geckoes are "perhaps the only example of a domesticated reptile," noting that "in towns in the tropics geckoes have . . . established themselves as house lizards."

Some Comcáac fishermen with whom I discussed it found Rosen's and Richardson's notion plausible, for they know that geckoes are often abundant in beach camps and villages, where they easily slip into bags, boxes, and other carryables. While only a few Seri fishermen reported having found geckoes in their boats, they agreed that they are highly mobile animals. If geckoes arrived on certain central Gulf coast islands by hitching rides with Seri or other fishermen, I estimate that such cultural dispersals would have affected the geographic distributions of over 10 percent of all fifty-some land reptile species occurring there.

THIS FOCUS on cultural dispersal of island species belies the vast majority of studies on island biogeography published since 1967, when a revolution began that shifted most ecologists' thinking about the patterns of life found on islands and their nearest continents. This revolution was inspired by a thin but challenging volume entitled *The Theory of Island Biogeography* (MacArthur and Wilson

1967), which proposed ways to determine whether the flora and fauna of particular islands have reached "equilibrium" with the area of habitat available to them. Its senior author was a brilliant, mathematically inclined avian ecologist, Robert MacArthur, who died a short time later. The "junior author," ant biologist Edward O. Wilson, who has since won two Pulitzer Prizes, was among the first conservation biologists to use equilibrium theory to predict the loss of biodiversity by habitat fragmentation (Quammen 1996; Mann and Plummer 1995). Soon afterward, Wilson and his colleague Robert W. Taylor made an attempt at incorporating thirty-five "tramp" species of ants—carried through the Polynesian archipelago by prehistoric or historic seafarers—into quantitative assessments of the equilibrium theory. Unfortunately, no other theorists followed up on their advance (Wilson and Taylor 1967; E. O. Wilson, pers. comm.).

One of the most widely cited biological treatises since Darwin's *Origin of Species*, *The Theory of Island Biogeography* triggered hundreds of new experiments and field studies on natural and simulated islands, bringing a new mathematical rigor to ecology and biogeography. At the same time, both followers and detractors of MacArthur and Wilson's equilibrium theory have shown that almost every principle set forth in that slim volume has been proven imprecise or at least inadequate (Mann and Plummer 1995). Although they inspired many advances, the initial formulas associated with equilibrium theory itself are now considered somewhat crude means of accounting for the diversity of life found on different kinds of islands, based on their sizes and distances from mainland sources of floral and faunistic richness (Williamson 1989).

Even at the time of its publication, geographer Jonathan Sauer (1969) wondered why MacArthur and Wilson's predictions of islands' species diversity varied so widely from what island biologists actually found in field studies. Sauer warned that the nascent theory was hopelessly ignorant of biological realities, treating as it did all species as equals, all modes of dispersal as equally probable, and all islands as homogeneous in their habitats. He wrote that "MacArthur and Wilson have, in effect, proposed that a gross and simplified version" of a quantitative theory replace the descriptive, empirically based field studies, which Sauer felt could better elucidate geographic patterns. Sauer continued his complaint: "In [MacArthur and Wilson's] equilibrium models, whole islands become sample quadrats in which the only data recorded are counts of species present. In terms of a data matrix in which the columns represent different sites and the rows represent different species, they lump all sites on an island into a single column and sum the species present in the column. All information on patterns within rows is thrown out with the arithmetical bath" (Sauer 1969).

Sauer sternly warned biogeographers not to succumb to what field biologists now refer to as "physics envy," rightly stating that however mathematically elegant such theories may be, they fall far short of true explanations of complex biological phenomena. He concluded: "The authors admit to certain weaknesses and crudities in their models, and the implication is that these can be remedied by tinkering and polishing. The defects seem to me to be more fundamental than that" (Sauer 1969).

Sauer is not alone in his assessment that MacArthur and Wilson's island equilibrium theory needed more than mere fine-tuning (Williamson 1989; Simberloff et al. 1992). Recently, biogeographer Robert J. Whittaker concluded that the theory is "essentially dead" (quoted in Hanski 1999). Although many biogeographers working today are still not ready to bury the theory six feet deep, they are painfully aware that it fails to predict much of the biogeographic patterning found in a loose archipelago such as the midriff islands of the Sea of Cortés (Simberloff et al. 1992).

In what was to become one of the most comprehensive tests of the equilibrium theory of island biogeography ever attempted across taxonomic lines for a single bioregion, quantitative biogeographers Ted Case and Martin Cody assembled two dozen scientists in Los Angeles in 1977 and asked them to use their detailed knowledge of island faunas in the Sea of Cortés to evaluate the theory's heuristic power. When their five hundred–some pages of analyses were finally published, Case and Cody (1983) conceded that "the work and results reported here show no unanimous support for the equilibrium theory."

One fatal weakness of the theory, they concluded, was its assumption that "islands are simple, small, and circumscribed chunks of the mainland which will provide habitats and resources like the mainland's but over a circumscribed area. The fallacy of this is illustrated by the small islands of the Sea of Cortez, which by intercepting a smaller total precipitation, fail to develop the topography and drainage systems that similar-sized pieces of the mainland, embedded in a contiguous landscape, have" (Case and Cody 1983).

The other fallacy of the theory—ironic, given that it was elaborated by two of the century's finest field ecologists—is that it ignores any biotic interactions an island plant or animal might have with other species, including humans. Case and Cody (1983, 1987) politely noted the weakness of this approach: "In short . . . the existence of an equilibrium species number [according to the theory] is a consequence of the simplest and most basic assumptions possible. It does not depend upon any biological interactions among the species. In fact, in the simplest case of linear in-migration and extinction curves, the implicit assumption is made that such interactions are absent."

Ted Case and other ecologists now fully reject that "implicit assumption," having convincingly demonstrated that interactions between species do indeed shape the biota of particular islands (Petren and Case 1997). This is especially true on small, recently emerged islands, ones formerly presumed to have so many open niches that neither competition nor cooperation among species would have much influence.

While island biogeographers deserve credit for broadening their repertoire of ecological principles to accommodate the effects of predator-prey and nurse plant–understory species interactions, few have found human-reptile interactions equally interesting. Fewer still have wondered whether such interactions can account for some of the variance from theoretical predictions for faunas found on particular islands in the Sea of Cortés. Even Conrad Bahre (1983), a cultural geographer and Seri historian recruited by Case and Cody to assess the human factor in their grand test of island biogeographic theory, failed to consider the extent to which the Comcáac were involved in faunal mixing. He simply mentioned that the Seri once made trips to San Esteban to collect chuckwallas for food—as if that were inconsequential to present biogeographic patterns.

It may now be possible to reach some new "consilience" of natural and human dispersal theories (Wilson 1988) to determine whether ancient human influences had any influence on the distributions of the sets of organisms that most dramatically deviated from the equilibrium theory's predictions. Even if human dispersal of fauna were too complex to factor into quantitative formulas, such dispersals could at least have been discussed *as part of the test* rather than as extraneous to it.

The Comcáac community is not well served by culturally naive approaches to biogeography, which ignore their historic influences on and, incidentally, their rights to certain biological resources. Geographer Jack Kloppenburg (1992) puts it this way: "What we call science enjoys a preeminent position among the possible ways of establishing knowledge about the world. Scientific rationality has achieved *de facto* status as the modern . . . standard against which all other knowledge claims are compared. Yet in many places, the constitution and character of existing science are being challenged as people come to recognize that the dominant mode of knowledge production does not necessarily serve their interests or meet their needs." In the case of the Comcáac, the elders who know of their culture's dispersal of certain reptile species were until recently not even asked to verify herpetologists' hypotheses regarding "aboriginal introductions." Nor did they have any reason to divulge such information to the outside world until they realized that by elucidating their influences, they would have a better chance of having Comcáac rights to local resources reaffirmed.

WHEN COMCÁAC POLITICAL LEADERS invited the Arizona-Sonora Desert Museum to help them build and manage a captive breeding area for San Esteban's Piebald Chuckwallas, I became party to lively discussions among Seri individuals who had different opinions on where the breeding pool of chuckwallas should be obtained. A few Seri men at first objected to the proposal that four chuckwallas be captured on Isla San Esteban and translocated to Punta Chueca because they felt that the political leaders had not consulted the traditional guardians of San Esteban and its resources. These guardians, they said, were elderly men related to the last survivor of the San Esteban group of Comcáac, who had regularly relied on the island's chuckwallas as a survival food during times of food stress.

As the entire community developed a broader sense of ownership of the project, this objection was withdrawn, but it was suggested that these guardians assist in determining the allowable size, age, and sex of animals to be translocated. It was further suggested that after a surplus of young Piebald Chuckwallas were produced, their parents should be taken back to San Esteban for release by the Comcáac guardians.

When I heard of this proposal to put captive chuckwallas back on the island, I pointed out that they might introduce diseases or mites from mainland wildlife to the island population. The Comcáac community understood my concern, and so we entertained other options.

I suggested that we obtain the chuckwallas not from San Esteban but from a small islet where biologists had clandestinely translocated a few individuals over a decade earlier (Hollingsworth et al. 1997). However, because this islet was managed as a protected area and was remote from Comcáac homelands, some government officials were uncomfortable with the prospect of Seri fishermen moving animals between localities.

The Comcáac responded by proposing to obtain the "San Esteban (Piebald) Chuckwallas" from Isla Alcatraz, an island closer to the coast where they had a more recent cultural legacy. The individuals who proposed this solution claimed that the chuckwallas on Alcatraz were identical to those on San Esteban in size and coloring.

For years I had heard from various herpetologists that the Isla Alcatraz chuckwallas were of hybrid origin (*Sauromalus varius* × *hispidus* × *obesus*) and so I was surprised that some Seri thought they and the San Esteban animals were identical. I should not have been surprised, however, for that was also the first impression of eminent herpetologists Chuck Lowe and Ken Norris (1955) after visiting the islands. I decided to see for myself.

In spring 1998 I worked with Craig Ivanyi and other Desert Museum herpetologists taking a series of measurements on the Alcatraz chuckwallas to compare with individuals of three species held captive at the museum (Fig. 8).

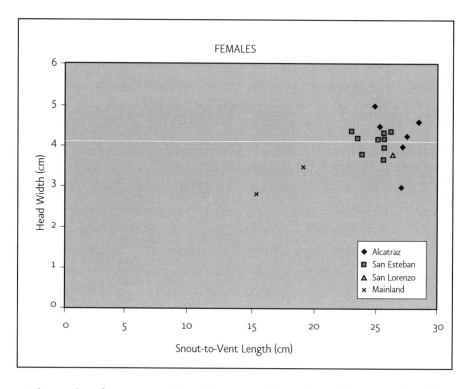

FEMALES

Figure 8. Relationship between head width and snout-to-vent length for female (*above*) and male (*right*) chuckwallas live-captured and released on Isla Alcatraz, compared with captive-bred chuckwallas at the Arizona-Sonora Desert Museum derived from San Esteban, San Lorenzo, and mainland populations.

The results of our census show that most of the individuals we sampled fall within a range of variation in size and color akin to that found among San Esteban Piebald Chuckwallas (*Sauromalus varius*) and San Lorenzo Black Chuckwallas (*S. hispidus*). However, a few individuals show constraints in size more frequently encountered among chuckwallas from the mainland (*S. obesus*). Our initial interpretation was that the progeny exhibiting traits of the more gigantic Piebald Chuckwallas numerically dominate this culturally initiated population, with secondary influences from Black Chuckwallas and only minor influences from the mainland populations closest to Isla Alcatraz. However, we cannot yet rule out the possibility that this population was derived from just two rather than three introductions. In any case, it is entirely plausible that the Comcáac created their own "breed" of somewhat tame, rather meaty chuckwallas as a backup food source in case they became stranded on the small island, thus taking the first steps in incipient domestication of faunal species.

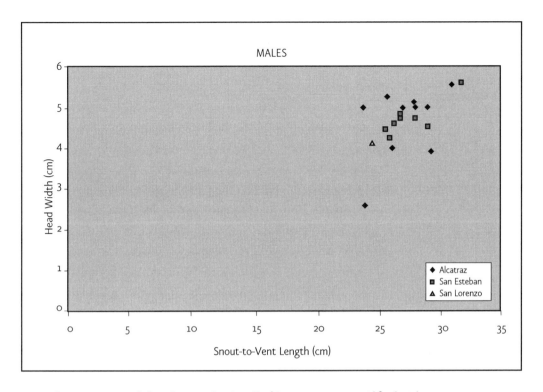

MALES

When I mentioned the Alcatraz chuckwallas' heterogeneity to Alfredo López, as it turned out, he told me that some Seri did not know the entire story: the buff, mottled Piebald Chuckwallas, he said, were brought to Alcatraz from San Esteban, but Black Chuckwallas were also transported there from San Lorenzo. He understood that these translocations occurred between 1925 and 1930, when Comcáac fishermen began to sell totoaba to fish buyers in Bahía Kino, making frequent forays to all the midriff islands as part of their operations (Bahre, Bourillón, and Torre 2000). Alfredo did not mention whether the island, less than a kilometer from the Sonoran mainland, already had a resident population of mainland chuckwallas or whether those, too, were brought there by Comcáac fishermen.

Bahre, Bourillón, and Torre place the onset of the commercial totoaba fishery boom in Kino closer to 1930, but agree that this industry ushered in a new era of mobility among the Comcáac. A few Seri fishermen maintained camps at Bahía Kino during off seasons, but they then had fewer food choices. And so they brought the gigantic chuckwallas from the other islands and established them on Alcatraz.

ON OUR FIRST BOAT TRIP passing Hastísil (Isla Cholludo) together, Alfredo said that the spiny-tailed iguanas there had been brought from San Esteban. He was struck by the fact that they were not brought to nearby Isla Dátil at the same time. Later, when I asked him if he knew of oral history verifying that the animals founding the San Esteban population were brought in from Isla San Pedro Nolasco—known in Cmique litom as *Hast Heepni Iti Ihom*, "Rock Where the Spiny-tailed Iguana Lies"—he admitted that he did not. Regardless, it may be that, prehistorically, this island near Guaymas was indeed the northernmost source of animals available to the Comcáac. This possibility is borne out to some degree by local speculations. Upon visiting San Pedro Nolasco with me and seeing an iguana carcass there, for example, Adolfo Burgos and Amalia Astorga suggested that long ago the mythic figure Hant Caai must have carried the iguanas from the island to other places where the Comcáac lived.

Such oral histories are not too far from the pattern of iguanid cultural dispersal that Lee Grismer (1994a) proposed:

> The distribution of the large iguanid lizards *Ctenosaura hemilopha* [now known as *C. conspicuosa*], *Sauromalus hispidus*, and *S. varius* on Isla Alcatraz, and *S. hispidus* on various islets within Bahía de los Ángeles or islands on which aboriginal inhabitants lived (Isla San Esteban) would facilitate their recapture as well as explain their rather anomalous distribution patterns.

We are left with two remaining puzzles, one regarding Desert Tortoises, the other regarding the Peninsular Leaf-toed Geckoes. Although Desert Tortoises are found on Isla Tiburón in abundance, they are not found on Islas San Esteban, Alcatraz, or Cholludo, and there is but one dubious report that they once occurred on Dátil. (In three visits there with herpetologists, I have been unable to confirm that Desert Tortoises persist on Dátil.) I know of no Comcáac oral history regarding the translocation of tortoises from one island to the next.

The other puzzle relates to the Peninsular Leaf-toed Gecko of the Sea of Cortés (*Phyllodactylus xanti*), which is widespread in Baja and its nearby islands but anomalous elsewhere. A sibling species is known from Ángel de la Guarda and the two San Lorenzos off central Baja California. The subspecies *P. xanti estebanensis* has been recorded on Islas San Esteban, Tiburón, and Alcatraz but to date has not been found on Islas Dátil and Cholludo, which are much closer to San Esteban and Tiburón than Alcatraz is.

It would be amazing if natural dispersal mechanisms took these geckoes to Isla Alcatraz from 25 kilometers or more to the west but did not take them just another kilometer, to the Sonoran mainland at Bahía Kino. And yet, the geckoes have been reported from Alcatraz but not on Bahía Kino's beaches, spitting distance away. Two other leaf-toed geckoes have spotty distributions on the mainland, with one found near Guaymas and another found just once north of Puerto Libertad. These anomalies may simply be the result of biogeographic squirreliness, as Ted Case calls it (Quammen 1996), but let us not forget that the hum of a Comcáac boat might be part of the environmental background noise.

The trouble is, most Comcáac do not like to touch, let alone think about or talk about, any of the geckoes or night lizards they collectively call *cozíxoj*. Those who did agree to discuss these psychologically dangerous creatures with me offered completely divergent views on how geckoes of any kind get around. Some women had been terrified by geckoes jumping off of tree branches at them. Others felt that the *cozíxoj* had supernatural powers that allowed them to actively swim or passively float on the sea's surface for long distances. Still others, though unconvinced that geckoes could ever enter a boat on their own, suggested that they could ride rafts of flood-driven driftwood down desert watercourses, into the sea, and out to the islands.

Above I suggested that the geckoes may have hitchhiked long ago, hidden in the reeds of kayaklike balsas or, more recently, in pangas and larger boats. And indeed, the explanation of cultural dispersal is more consistent with their current distribution than is natural long-distance dispersal by drifting, swimming, or floating. This stowaway hypothesis for gecko dispersal is supported by the fact that most Comcáac adults dread finding a gecko anywhere in their vicinity. If they do discover one in their home, they become obsessed with removing it. The same would likely hold true if they found one had traveled with them to another camp or, conceivably, to another island—and so the little animal would perforce have found a new home. (Herpetologist Borys Malkin reported a similar apprehension about geckoes in South America, and Maurice Richardson [1972] observed that "in some parts of Africa they are feared and persecuted.")

As we have seen, the stowaway hypothesis is well accepted by Hawaiian herpetologists (Fisher 1997; McKeown 1978). Recently, evolutionary biologist Chris Austin (1999) used the degree of genetic divergence among various islands' "stowaway" lizards, which were presumably dispersed by the first seafaring colonists in Polynesia, to examine the rate of early human migration in the central and eastern Pacific. It may be possible to do the same with geckoes on the midriff islands, testing whether early Hokan speakers inadvertently carried geckoes as they island-hopped from Baja California to Sonora to "become" the Seri, or whether they moved in ancient times by land from the Californias to Sonora.

There is a message here for all of us: sometimes we are attractive even when we don't want to be. Seri households, for example, offer valuable microhabitats for numerous terrestrial reptiles, which are attracted to the food and water, shade, and shelter provided by human constructions. A Desert Museum–sponsored inventory of horticultural introductions to Comcáac dooryard gardens within the last two decades lists more than sixty species of vegetables and landscape plants currently present in the two villages (Tables 10 and 11). These plants offer desert reptiles more diverse food and shelter options than they ever experienced along the central Gulf coast prior to 1974.

Although they do not embrace an agricultural tradition per se, the Comcáac have likely been moving plants around for may decades. Felger and Moser (1985) noted disjunct populations of prickly pears (*Opuntia phaecantha* var. *discata*) at Punta Sargento and on Isla Alcatraz, which were clearly planted there. Yetman and Burquez (1996) proposed that the Comcáac have historically translocated two

TABLE 10

Cultivated Plants Found within Desemboque Village, 1996–2000

Species	Management	Arizona-Sonora Desert Museum Voucher
AGAVACEAE		
Agave colorata	Cultivated	No specimen
A. pelona	Cultivated	No specimen
AIZOACEAE		
Caprobrotus angula	Cultivated	ASDM 8.98.9
Drosantheum sp.	Cultivated	ASDM 8.98.12
Sesuvium verrucosum	Self-seeding	ASDM 8.98.3
ANACARDIACEAE		
Rhus sp.	Cultivated	ASDM 8.98.17
ARECACEAE		
Phoenix dactylifera	Cultivated	No specimen
ASTERACEAE		
Ambrosia ambrosioides	Self-seeding	No specimen
A. confertifolia	Self-seeding	ASDM 8.98.24
Artemesia sp.	Cultivated?	No specimen
Baccharis salicifolia	Self-seeding	ASDM 8.98.27
Helianthus annuus	Cultivated	ASDM 8.98.13
BIGNONIACEAE		
Spathodea sp.	Cultivated	No specimen
Tecoma stans	Cultivated	No specimen
BORAGINACEAE		
Cryptantha maritima	Self-seeding	ASDM 8.96.6
BURSERACEAE		
Bursera sp.	Cultivated	No specimen
CACTACEAE		
Carnegiea gigantea	Cultivated	No specimen
Opuntia ficus-indica	Cultivated	No specimen
Pachycereus pringlei	Cultivated	No specimen
Stenocereus gummosus	Cultivated?	No specimen
S. schottii	Cultivated	No specimen
CANNACEAE		
Canna flaccida	Cultivated	No specimen
CHENOPODIACEAE		
Atriplex barclayana	Self-seeding	ASDM 8.98.5
A. linearis	Self-seeding	ASDM 8.98.4
Suaeda moquinii	Self-seeding	ASDM 8.98.28

Continued on next page

TABLE 10 (continued)

Species	Management	Arizona-Sonora Desert Museum Voucher
CONVOLVULACEAE		
Merremia dissecta	Cultivated	ASDM 8.98.16
EUPHORBIACEAE		
Nerium oleander	Cultivated	No specimen
Ricinus communis	Cultivated	ASDM 8.98.15
FABACEAE		
Leucaena leucocephalum	Cultivated	ASDM 8.98.21
Parkinsonia aculeata	Cultivated	ASDM 8.98.20
P. florida	Self-seeding?	No specimen
Pithecellobium dulce	Cultivated	ASDM 8.98.19
Prosipasdastrum sp.	Cultivated	ASDM 8.98.18
Prosopis glandulosa	Cultivated	ASDM 8.98.29
Psorothamnus emoryi	Self-seeding	ASDM 8.98.1
FOUQUIERIACEAE		
Fouquieria splendens	Cultivated as fence	No specimen
FRANKENIACEAE		
Frankenia palmeri	Self-seeding	ASDM 8.98.2
LAMIACEAE		
Mentha sp.	Cultivated	No specimen
LILIACEAE		
Aloe barbadensis	Cultivated	ASDM 8.98.14
MALVACEAE		
Bakeridesia notophium	Cultivated	ASDM 8.98.25
MORACEAE		
Ficus benjamina	Cultivated	ASDM 8.98.23
MUSACEAE		
Musa paradisica	Cultivated	No specimen
MYRTACEAE		
Eucalyptus sp.	Cultivated	No specimen
NYCTAGINACEAE		
Bougainvillea sp.	Cultivated	No specimen
POACEAE		
Arundo donax	Cultivated	No specimen
RHAMNACEAE		
Rhamnus sp.	Cultivated	ASDM 8.98.26
ROSACEAE		
Rosa sp.	Cultivated	No specimen

TABLE 10 (continued)

Species	Management	Arizona-Sonora Desert Museum Voucher
RUTACEAE		
Citrus sp.	Cultivated	No specimen
SALICACEAE		
Populus fremontii	Cultivated	No specimen
SOLANACEAE		
Nicotiana glauca	Self-seeding	ASDM 8.98.8
TAMARICACEAE		
Tamarix aphylla	Cultivated	No specimen
VERBENACEAE		
Vitex mollis	Cultivated	ASDM 8.98.30

TABLE 11

Cultivated Plants Found within Punta Chueca Village, 1996–2000

Species	Management	Arizona-Sonora Desert Museum Voucher
AGAVACEAE		
Agave americana	Cultivated	No specimen
AIZOACEAE		
Trianthema portulacastrum	Self-seeding	No specimen
AMARANTHACEAE		
Amaranthus fimbriatus	Self-seeding	No specimen
A. palmeria	Self-seeding	No specimen
ANACARDIACEAE		
Rhus sp.	Cultivated	ASDM 8.98.39
APOCYNACEAE		
Huernia sp.	Cultivated	ASDM 8.98.46
Vinca major	Cultivated	No specimen
ARECACEAE		
Phoenix dactylifera	Cultivated	ASDM 8.98.42
ASTERACEAE		
Ambrosia ambrosioides	Self-seeding	No specimen
Tagetes sp.	Cultivated	No specimen
Trixis californica	Cultivated	No specimen
Verbesina enceliodoides	Cultivated?	ASDM 8.98.35
BIGNONIACEAE		
Tecoma stans	Cultivated	No specimen

Continued on next page

TABLE 11 (continued)

Species	Management	Arizona-Sonora Desert Museum Voucher
CACTACEAE		
Carnegiea gigantea	Cultivated	No specimen
Ferocactus sp.	Cultivated	No specimen
Opuntia cylindracea	Cultivated?	No specimen
O. ficus-indica	Cultivated	No specimen
O. violacea	Cultivated	No specimen
Pachycereus pringlei	Cultivated	No specimen
Stenocereus alamosensis	Cultivated	No specimen
S. thurberi	Cultivated	No specimen
CUCURBITACEAE		
Citrullus lanatus	Cultivated	No specimen
EUPHORBIACEAE		
Ricinus communis	Cultivated	ASDM 8.98.15
FABACEAE		
Caesalpinia sp.	Cultivated	ASDM 8.98.34
Olneya tesota	Self-seeding	No specimen
Palafoxia arida	Self-seeding	ASDM 8.98.41
Parkinsonia aculeata	Self-seeding	No specimen
Prosopis glandulosa	Cultivated?	No specimen
LAMIACEAE		
Hyptis albida	Cultivated	ASDM 8.98.38
H. emoryi	Self-seeding?	No specimen
Ocimum basilicum	Cultivated	ASDM 8.98.43
LILIACEAE		
Aloe barbadensis	Cultivated	No specimen
Allium sp.	Cultivated	No specimen
MORACEAE		
Cannabis sativa	Cultivated	No specimen
Ficus benjamina	Cultivated	No specimen
MYRTACEAE		
Eucalyptus sp.	Cultivated	No specimen
NYCTAGINACEAE		
Boerhaavia coccinea	Self-seeding	ASDM 8.98.36
Bougainvillea sp.	Cultivated	ASDM 8.98.36
POACEAE		
Arundo donax	Cultivated	No specimen
PORTULACACEAE		
Portulaca oleracea	Cultivated	ASDM 8.98.47

TABLE 11 (continued)

Species	Management	Arizona-Sonora Desert Museum Voucher
SAPINDACEAE		
Cardiospermum corindum	Cultivated	ASDM 8.98.32
SOLANACEAE		
Capsicum annuum	Cultivated	ASDM 8.98.44
Lycopersicon esculentum	Cultivated	No specimen
STERCULIACEAE		
Melochia tomentosa	Cultivated	ASDM 8.98.33
VERBENACEAE		
Vitex sp.	Cultivated	ASDM 8.98.37
ZYGOPHYLLACEAE		
Tribulus terrestris	Self-seeding	ASDM 8.98.45

Note: Special thanks to Barb Skye, Patty West, David Seibert, Kim Buck, and Jim Donovan for helping collect these data.

cactuses with edible fruit to their Cerro Prieto refuge, the Cardón (*Pachycereus pringlei*) and Pitahaya Agria (*Stenocereus gummosus*). Neither of these studies, however, proposed how the Comcáac might have dispersed these cactuses, or when.

Interviews I have conducted with Comcáac adults confirm that their families have a tradition of transporting and transplanting small specimens of certain highly prized perennials, including Cardón, Pitahaya Agria, Pitahaya Dulce or Organpipe (*Stenocereus thurberi*), Saguaro (*Carnegiea gigantea*), Prickly Pear (*Opuntia engelmanii*), elephant trees (*Bursera* spp.), and mesquites (*Prosopis* spp.). This transplanting process may also account for some of the populations of Organpipe found by Bowen on the northwest corner of San Esteban (Felger and Moser 1985; Bowen 2000), one of them a ring of even-aged cacti apparently planted around an ancient camp (Nabhan 2000c).

"Yes, it's possible that they moved cactus from one camp to another to mark a site, to offer a sign that it is a birthing ground. [Your family] can also mark your birthplace with stones, or bury your placenta under a Saguaro, an Organpipe, a Prickly

Pear, or a Cardón. . . . Here [showing me a 25- to 40-centimeter-tall Organpipe],
this is the size they carried so they wouldn't dry out."

ALFREDO LÓPEZ, *on a field trip into the hills inland from Punta Chueca*

When I asked why these perennials were transplanted, I was told that the parents, in-laws, or godparents had an obligation to show a child, as he or she grew up, where the newborn's iixöni, "placenta," was buried—in essence, where the child was connected to the earth. If such people were residing away from where the pregnant mother went into labor, a runner would go to notify them, and they would bring a small plant with them to commemorate the nativity site. A hardy, long-lived plant propagated above the buried placenta made an excellent living landmark, one that would grow just as the child would (Nabhan 2000c).

"Yes, our Comcáac ancestors [*Seris antiguos*] transplanted cactus. . . . Maybe someone brought this Pitahaya Agria here to Tecomate from Xapoo Ilitcoj [several kilometers away]. There are many others elsewhere, in particular places. . . . They carried cactus. . . . When a little child is born, they put this stick, they nailed it in, to see afterward. The plant then has an owner who was born the same day it was transplanted. Yes, that is where they put the placenta, as I said."

AMALIA ASTORGA, *on Comcáac cactus-transplanting traditions*

Human dispersal continues to reshape reptile habitats on the mainland as well as on the islands. The Comcáac name several reptiles that they claim are more abundant inside their villages than on surrounding scrublands: Western Whiptails, banded geckoes, Tree Lizards, and Zebra-tailed Lizards. Other reptiles—Regal Horned Lizards, Desert Spiny Lizards, Gopher Snakes, sandsnakes, and various rattlers, including Sidewinders—come into both Punta Chueca and Desemboque as well, even though they may never become abundant in houseyards. Still other reptiles, however, such as Desert Tortoises, Gila Monsters, Desert Iguanas, and collared lizards, actually seem to avoid their villages.

These perceptions are interesting and form the basis for a more detailed inquiry that is just beginning, assessing Comcáac villages as reptile habitats. During the summer of 1998 several biologists helped me conduct walking surveys to establish the identity and abundance of reptiles inside Punta Chueca and Desemboque and in nearby areas of similar size and habitat type. We also invited Comcáac children to live-capture and slingshot-stun any reptiles they could find inside the villages but outside our survey plots.

Our preliminary results illustrate the obvious: (1) children more easily catch lizards than adults do; and (2) contemporary Comcáac villages serve as artificial oases for some native species, but they repel others. Despite the fact that villages offer permanent sources of water, structural heterogeneity for roosts, and plenty of artificial crevices for nesting or escape, they also pose significant threats to reptilian well-being: fast cars, carnivorous pets, parasites, and microbes. Only Western Whiptails and Side-blotched Lizards seem to thrive in the presence of such perils, while Tree Lizards, banded geckoes, and Zebra-tailed Lizards find these new habitats to be both a blessing and a curse. The risks in Comcáac villages today are vastly different from those of the historic encampments, where a small number of families stayed for only a few weeks and so had a relatively short-term impact on lizards' and snakes' habitat.

Perhaps implicit in the Comcáac concept of *iime* as "refuge" or "nesting grounds" is that some animals are better off taking refuge away from humans than staying with them and their pets. Nevertheless, some animals do take refuge with the Comcáac, and the Comcáac in turn have experimented with moving species they consider culturally significant into their village communities. While habitat modifications and transport of certain reptiles have made for a few distributions that initially appeared "squirrelly" to outsiders, the long-term effects of these activities on zoogeography now appear to be explainable in terms that both Western scientists and Seri seafarers can understand. Wherever the Comcáac congregate, there are bound to be a few tolerated camp followers, a few tramps, a few intentional introductions, and a few dangerous visitors who arrive on their own, even when a welcome mat is not set out for them.

naming the menagerie

How to Sort One Snake from Another

The Mexican sierra [a Sea of Cortez fish] has "XVII-15-IX" spines in the dorsal fin. These can easily be counted. But if the sierra strikes hard on the line so that our hands are burned, if the fish sounds and nearly escapes, and finally comes in over the rail, his colors pulsing and his tail beating in the air, a whole new relational externality has come into being—an entity which is more than the sum of the fish plus the fisherman. The only way to count the spines of the sierra unaffected by this second relational reality is to sit in a laboratory, open an evil-smelling jar, remove a stiff colorless fish from formalin solution, count the spines and write the truth, "D.XVII-15-IX." There, you have recorded a reality, which cannot be assailed—probably the least important reality concerning either the fish or yourself.

JOHN STEINBECK AND EDWARD F. RICKETTS, *The Sea of Cortez* (1941)

We were lifting our waterproof bags of belongings out of the panga
to set up a Thanksgiving camp at Ensenada del Perro on Isla Tiburón.
As my two Seri companions waded armloads of bags to shore in front of
me, two or three Zebra-tailed Lizards lifted their scorpionlike tails above
the sandy beach where they had been resting, wagged them at us, and
ran off into the bushes.

"*Ctamófi*," mumbled one of the Seri fishermen to the other. The other
grunted back affirmatively, not even stopping to give the lizards a good
look. Less than two seconds had passed since he turned from the boat,
glimpsed the tails of the lizards out of the corner of his eye, and seen the
animals disappear.

"So they do call them *ctamófi!*" I said to myself, amazed to hear a Seri
make a positive field identification of this species so rapidly. On my last
visit to their village, I had shown photos, drawings, and even pickled
specimens of them to many different Seri individuals, but it seemed that
with each inquiry I was given a different name for this common little
lizard. I had become so frustrated that I had quit asking about it. My
colleague Janice Rosenberg had suffered similar results the spring before;

only 59 percent of Comcáac adults over forty summoned up the name *ctamófi* when shown a color photo of its profile, while the other 41 percent offered a wide scatter of other names. None of the children, who were even more confused by the lizard image in the photo, offered the name *ctamófi*.

As I plopped my bags down on the beach, it dawned on me: those side-view photos did not capture the scorpionlike tail wag, or the stripes on the tail's underside, which are seen only when it flashes a warning to a potential predator. The limp, pickled specimens did no better, neither revealing the dramatic upcurved tail nor conjuring up its waggle-dance warning.

And yet, my Seri friends needed only a split-second glance at this creature performing its characteristic behavior in its natural habitat to bring its name to their lips.

That is how I came upon the Zebra-tailed Rule of Field Taxonomy: a herp on the run is worth two dead in the hand. And yet, even a simple rule like that is not foolproof. About a year later, Craig Ivanyi and I took Guadalupe López with us to Punta Santa Rosa on a foraging trip. When Craig captured a Zebra-tailed Lizard, I was sure that Guadalupe would look at it and declare it to be a *ctamófi*. No such luck.

"*Ctoicj*," he declared.

"*Ctoicj*? Why not *ctamófi*?" I moaned.

"Yes, well . . . it's not easy to know the names because, traditionally, the animals of the *monte* all had two, maybe three names," he explained. "*Ctoicj* is the oldest one, but it is for the same animal as *ctamófi*." He added, "There is a third name for young ones: *matpayóo*."

I groaned. Then he explained to me a similar naming situation for whiptail lizards. I groaned again.

"Don't worry, Hant Coáaxoj, most of them have two or three names, and at least you know one of them."

BEFORE THE EARTH was formed there were already many kinds of marine animals, the giant sea turtle among them. When Turtle brought mud from the bottom of the sea so that Hant Caai could make the land, Hant Caai used the last bit left on Turtle's flipper to shape the first humans. After the land was made, *Hant Iha Quimx,* "Earthspeaker," began to teach humans the names of all the animals. He attempted to do so according to some rather simple rules.

In any culture, different people interpret such rules in their own ways. As the Zebra-tailed Lizard example demonstrates, a member of the Comcáac community may use different names at different times for the same reptile. Indeed, the variety of names I have learned for the same animal is striking, as is the multiplicity of diagnostic criteria different Seri teachers have taught me for identifying reptiles in their homeland.

According to Comcáac tradition, animals are collectively called *ziix ccam,* "thing that-is-alive" (Fig. 9). This term, called a "unique beginner" by scholars who study folk taxonomies, includes within its domain a number of life-forms which the Comcáac recognize by their gross morphology and mode of locomotion. Among these life-forms are *zixcám,* "fish," a name similar to that for all living things, perhaps because they were among the first mythically recognized during Creation. Other such terms include *ziic,* "bird," and *xica ccam heecot cocom,* "things-that-are-alive lying-in-a-desert-area." According to Becky Moser, only a few Seri know this last life-form category, using it to refer especially to four-legged fur bearers.

There is another life-form term that includes most four-legged terrestrial reptiles, *ziix haquímet hant cöquiih,* "thing lizard land what-is-on." A kindred category includes venomous snakes, *ziix cocázni hant cöquiih,* "thing biting land what-is-on." Still another describes snakes with nonvenomous bites, *ziix coimaj hant cöquiih,* "thing serpent land what-is-on." At least one other category for an animal life-form is commonly used; it includes marine and freshwater turtles as well as Desert Tortoises: *ziix moosni hant cöquiih,* "thing turtle land what-is-on" (Fig. 10).

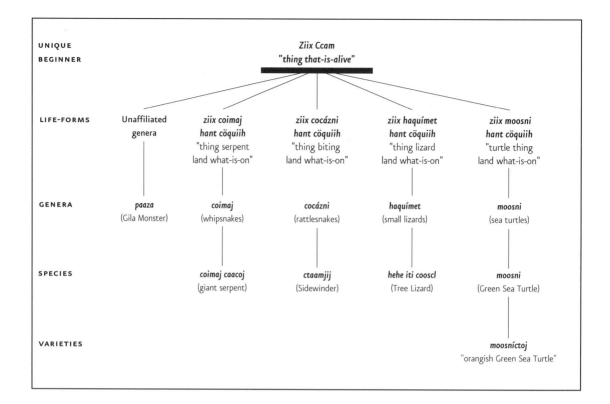

UNIQUE BEGINNER		*Ziix Ccam* *"thing that-is-alive"*			
LIFE-FORMS	Unaffiliated genera	*ziix coimaj* *hant cöquiih* "thing serpent land what-is-on"	*ziix cocázni* *hant cöquiih* "thing biting land what-is-on"	*ziix haquímet* *hant cöquiih* "thing lizard land what-is-on"	*ziix moosni* *hant cöquiih* "turtle thing land what-is-on"
GENERA	*paaza* (Gila Monster)	*coimaj* (whipsnakes)	*cocázni* (rattlesnakes)	*haquímet* (small lizards)	*moosni* (sea turtles)
SPECIES		*coimaj caacoj* (giant serpent)	*ctaamjij* (Sidewinder)	*hehe iti cooscl* (Tree Lizard)	*moosni* (Green Sea Turtle)
VARIETIES					*moosníctoj* "orangish Green Sea Turtle"

Figure 9 (*above*). Some life-form categories named in Cmique Iitom.

Figure 10 (*right*). A Desert Tortoise carving showing the characteristic shape of the carapace. (Photo Helga Teiwes, 1997)

With regard to other, more unique-looking, reptiles, some Seri individuals are so intimately familiar with them that they do not place them in any life-form category. These "unaffiliated genera" include *paaza*, Gila Monsters, and *cozíxoj*, geckoes and night lizards. Which lizards stand alone and which get placed within the life-form category *ziix haquímet hant cöquiih* is to some extent a matter of personal taste and familiarity. However, the number of monotypic, unaffiliated genera recognized by most Comcáac is quite large compared to the number of polytypic genera subsumed under life-form categories (Table 12). Scholars of folk taxonomies typically refer to a classification system as "flat" if it has many monotypic, unaffiliated genera and tends not to use life-forms and other higher categories. By these standards, Comcáac folk taxonomies are considered to be extremely flat (Berlin 1992). This characterization, however, does not do justice to the complexity of the Comcáac sense of natural order.

Oddly, a few kinds of animals and plants—humans, Leatherback Turtles, Regal Horned Lizards, and boojums, for example—are separated out from the "unique beginner" category *ziix ccam*. Instead, these legendary beings are placed in another, higher, category: *ziix quiisax*, "things with-spirit-breath" (Felger and Moser 1985).

"The Seri can't eat Leatherbacks. It's prohibited. If you put it [Leatherback meat] into your mouth, you'll fall under a spell like you're sleeping, even though you were just awake, chewing. The Leatherbacks understand our dialect, they can turn themselves over for us in the water or on the beach. It's not merely a turtle—it's one of us. Leatherbacks always come in the spring, migrating with fleets of *cooyam* [Green Sea Turtles from the south]. They never come alone, by themselves."

GUADALUPE LÓPEZ, *Desemboque*

EVEN IN THE DRIEST PORTIONS of the Sonoran Desert, myriad kinds of animals walk on the ground, protecting themselves with a shell or a venomous bite; they are not all easy to tell apart at first glance, so there are many names and diagnostic characters to remember. Sometimes elders in Desemboque will recall

TABLE 12

Polytypic and Monotypic Genera of Reptiles Recognized by the Comcáac

Polytypic	Western Scientific Equivalent
Moosni	*Gopherus agassizii* (Desert Tortoise), *Kinosternon* (mud turtles), and *Cheloniidae* (sea turtles)
Coof	*Sauromalus* (chuckwallas)
Ctoixa	*Cnemidophorus* (whiptail lizards)
Coimaj, maxáa	*Masticophis* (whipsnakes)
Coftj	*Micruroides* (coralsnakes)
Cocázni	*Crotalus* (rattlesnakes, including Sidewinders) and possibly other venomous snakes such as *Pelamis platurus* (Yellow-bellied Seasnake)

Monotypic (Unaffiliated?)	
Xtamáaija?	*Kinosternon* (mud turtles; see *moosni*, above)
Hast coof*	*Crotaphytus* (collared lizards)
Hantpízl*	*Gambelia wislizenii* (Long-nosed Leopard Lizard)
Heepni	*Ctenosaura* (spiny-tailed iguanas)
Meyo	*Dipsosaurus* (desert iguanas)
Ctamófi*?	*Callisaurus draconoides* (Zebra-tailed Lizard)
Hant coáaxoj	*Phrynosoma solare* (Regal Horned Lizard)
Hant caxaat	*Holbrookia maculata* (Lesser Earless Lizard)
Haasj	*Sceloporus* (spiny lizards)
Hehe iti cooscl*?	*Urosaurus* (brush lizards)
Yax quiip, tozípla, ctoicj, matpayóo	*Uta stansburiana* (Side-blotched Lizard)
Cozíxoj	*Coleonyx* and *Phyllodactylus* (geckoes), *Xantusia* (night lizards)
Paaza	*Heloderma suspectum* (Gila Monster)
Xasáacoj	*Boa constrictor* (Boa Constrictor)
Ziix haas ano cocázni	*Charina trivirgata* (Rosy Boa), *Trimorphodon biscutatus* (Lyresnake)
Cocaznáacöl	*Arizona elegans* (Glossy Snake), *Pituophis melanoleucus* (Gopher Snake), *Salvadora hexalepis* (Western Patch-nosed Snake)
Hapéquet camízj	*Chilomeniscus* (sandsnakes)
Hamísj catójoj	*Oxybelis aeneus* (Brown Vinesnake)
Xepe ano cocázni	*Pelamis platurus* (Yellow-bellied Seasnake)
Xepe ano paaza	*Crocodylus acutus* (River Crocodile)

*May be members of a larger grouping of small lizards, *haquímet*.
? = possibly affiliated.

a different name for an animal than the one those in Punta Chueca know it by. When Comcáac hunters were more isolated from one another and ranged more widely in their pursuit of game, this plurality of names may have been even greater.

Nevertheless, Hant Iha Quimx must have begun with a certain logic for shaping the names of particular creatures so that the Comcáac could remember how one was different from another. Today, though this logic has drifted this way and that, it can still be recognized in the structure of animal names.

Consider the chuckwalla, for example. During a time when the Comcáac knew just one kind of chuckwalla within their prehistoric territory, it was apparently called simply *coof*, polysemous with a verb, "to hiss" or "to blow air." Later, when they encountered a second kind of chuckwalla on Isla San Lorenzo or on Ángel de la Guarda, they called this large, dark-colored species *coof coopol*, or "chuckwalla blackish." However, when they came upon another kind of reptile in the mountains that reminded them of chuckwallas, they fashioned a name for it using a modifier in front of the generic term for chuckwalla: *hast coof*, for instance, a name that means something like "chuckwalla-like mountain dweller" and is now used for collared lizards. If it were possible for a Seri to make the construction *coof hast*, it might mean something closer to "mountain made of chuckwallas."

"*Hast coof*—that's a pretty one, full of color. It's here on the mainland, in the mountains. When you throw a rock at it, it won't run away from you, you can't force it from its place. It will hold its ground or pretend like it will attack you."
MANUEL FLORES, *translated by Ignacio "Nacho" Barnet, Punta Chueca*

Following the guidelines implicit in the names offered by Hant Iha Quimx, Comcáac children knew which animals were specific kinds of chuckwallas and which were chuckwalla-like animals of a different sort. However, not everyone stuck with the names that Hant Iha Quimx offered. Sometimes they would nickname a person for an animal, but when that person died they would retire the animal's original name and coin a descriptive name for it instead. Or they might

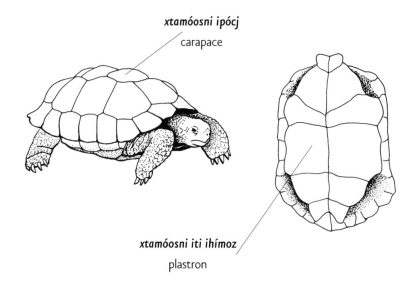

Figure 11. Desert Tortoise body parts named in Cmique Iitom.

xtamóosni ipócj
carapace

xtamóosni iti ihímoz
plastron

simply tire of the original name and come up with a riddlelike nickname for the animal. For instance, the chuckwallas on the mainland are now commonly described as *ziix hast iizx ano coom*, "thing that engulfs itself in a shelter of rocks."

Similarly, although Desert Tortoises were originally called *xtamóosni*, almost no one today uses that name except when singing or when referring to particular plants eaten by tortoises. Most Comcáac individuals interchangeably use two descriptive coinages: *ziix hehet cöquiij*, "thing sitting under plants," or *ziix catotim*, "thing that slowly scoots along." For such riddlelike coinages to become so common that they eclipse the original name, the creature's appearance and habits must be utterly well known (Nelson 1983). And indeed, virtually all Comcáac individuals know Desert Tortoises by their characteristic shape (Fig. 11; and see Fig. 10), and can readily sex them on the basis of sexual dimorphisms.

Of course, some reptiles are not seen as frequently as Desert Tortoises are and perhaps as a consequence have an ambiguous nature. For example, it has been a long time since the Comcáac have seen a crocodile outside of a zoo, and they do not know whether to call it *xepe ano heepni*, "spiny-tailed iguana of the sea," or *xepe ano paaza*, "Gila Monster of the sea." Some Seri individuals claim that only one of these names refers to River Crocodiles, while the other refers to a different marine creature altogether.

"Xepe ano paaza—it was here long before, perhaps during the era of my father's youth. It was seen in open water, once near Punta Sargento [on the mainland] and once at Casíime [on Isla Tiburón]. Its little ones lived in the estuaries, but the adults swam out from the coast. Today, there are too many long-lines and drift nets for it to survive. . . . I've only seen one once, when Santiago Loco took me to the zoo in Phoenix. Someone told us that what we were looking at was called a caiman. We talked among ourselves for a while, looking at it, and decided that it's what the old people called *xepe ano paaza!"*

ADOLFO BURGOS, *Desemboque*

The Yellow-bellied Seasnake, seen by Seri fishermen nearly as infrequently as River Crocodiles, is called *xepe ano cocázni*, "venomous snake of the sea." Alfredo López claims that it earned this name because it does have a venomous bite, even though it does not have the characteristic rattle that most *cocázni* have. In line with this reasoning, he places it in the life-form category *ziix cocázni hant cöquiih*, which also includes several color morphs of rattlesnakes (in five species recognized by Western scientists), as well as *ctaamjij*, "Sidewinder," and *coftj*, "coralsnake."

A few elderly Seri individuals distinguish the Tiger Rattlesnake from others by the name *cocázni cahtxíma*, but most do not. On the other hand, nearly all Comcáac recognize Sidewinders (Figs. 12 and 13) as a subset of rattlers, regardless of whether they ever call them by the rarely used term *cocázni ctaamjij*. Etymologically, this compound secondary lexeme may imply either that it is a horned rattler or that it is from a sandy habitat. Most Comcáac simply say *ctaamjij* when they see a Sidewinder.

In Alfredo's mind, the lack of a *yeesc*, "rattle," does not keep seasnakes or coralsnakes from being akin to rattlers. Perhaps this is because he has seen true *cocázni* on the islands which have deformed rattles that eventually fall off; they can give

Figure 12. A Sidewinder carved out of ironwood by José Luis Blanco. Note the way the curvilinear shape suggests slithering over sand. (Photo Helga Teiwes, 1997)

you a deadly bite even if you don't hear a warning rattle. In fact, venomous rattlesnakes on several of the Gulf islands have lost their rattles through genetic deformations, so this trait is not consistently useful in alerting one to impending peril. It is more critical that children learn to think of all members of the category *ziix cocázni hant cöquiih* as venomous, rattle or no.

Perhaps that explains why the Comcáac do not teach their children to differentiate Western Diamondbacks from Mohave Rattlers, a feat that even some herpetologists find difficult given the color variants of both species in the central Gulf coast of Sonora. The simple characteristic used to distinguish Diamondbacks and Mohaves elsewhere—that Mohaves have white tail bands that are wider than the black rings adjacent to them—simply does not work on the Sonoran coast or on Isla Tiburón. It is enough to know that both are venomous and simply get out of the way; no need to spend time looking at the width of tail stripes near the rattle when it's a matter of life and death.

It is curious that most contemporary Comcáac do not concern themselves with the differences between the true Western Coralsnake (*coftj*) and false coral-

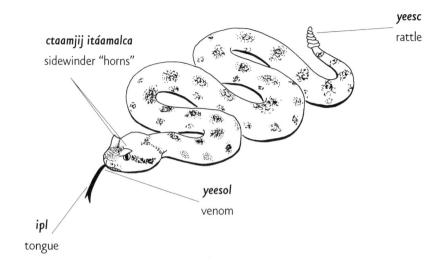

Figure 13. Sidewinder body parts named in Cmique litom.

yeesc
rattle

ctaamjij itáamalca
sidewinder "horns"

yeesol
venom

ipl
tongue

snake (*hant quip*). Though known to be venomous, the former is generally considered by the Comcáac to be as harmless as the false coralsnakes. Since the time of Borys Malkin's ethnozoological interviews among the Comcáac in the early 1950s, Seri individuals have often claimed that true coralsnake bites can be fatal to other animals or to the Cocsar (Mexicans) but that these snakes have never fatally poisoned a Seri person who has handled them. The Comcáac do distinguish the two snakes, as evidenced by a tongue twister labeling the sequence of colors found on the back of a true coralsnake. It is not all that different from the English rhyme, "Red and yellow, kill a fellow; red and white, he's all right." Perhaps part of the lack of concern stems from the fact that within Comcáac territory true coralsnakes are different enough in size and habit to be easily distinguished from *coralillos falsos* in most field situations, and so people simply steer clear of them.

In general, the Comcáac taxonomy of reptiles has a considerable degree of correspondence with Western (Linnaean) taxonomy. Some Western scientists, however, have argued that their own taxonomy of reptiles is more accurate than that of the Comcáac. Consider two commentaries written by Borys Malkin (1962) in the 1950s, which now appear rather embarrassing:

> Moreno [Malkin's Seri guide] told me that there are two species of *Dipsosaurus* [sic, iguanids] on San Esteban and the one I had was the smaller of the two. I am unable to check this information and it may be that what he had in mind was the larger chuckawalla [*Sauromalus varius*]. I have obtained information about the chuckawalla from other informants but not from him. There are no larger lizards in the area other than the chuckawalla and the Gila monster; therefore this suspicion [that José Juan Moreno is wrong] is justified.

Clearly, Malkin was not aware of the other large iguanid on Isla San Esteban, *Ctenosaura conspicuosa*, which, like the Desert Iguana (*Dipsosaurus dorsalis*), has small crests and dewlaps (Fig. 14). From my trips with Moreno to the islands, it is apparent that he readily makes a distinction between these two genera. They, along with chuckwallas, are grouped together in Comcáac tradition because the males of all three have enlarged, oozing pores on their femoral ridges (Fig. 15). (These pores are the subject of a Comcáac legend noted earlier in this volume.) In this case, Moreno was not confused; Malkin simply lacked the requisite familiarity with the fauna of the midriff islands to recognize the difference.

Malkin (1962) did admit, however, that "in distinguishing between one species and another, the Seris showed remarkable acuity. Out of 26 species listed here, they confused only two, both relatively small and similar *Utas*. On the other hand, they were able to associate correctly specimens of different sizes and often also the colors of the same species. I frequently mixed up a large number of specimens without telling them anything about them and yet invariably the informants were able to sort them correctly."

Ironically, of the two species that Malkin said the Comcáac confused, one is no longer considered to be an *Uta*. The Seri men I have been in the field with sometimes confuse this species, a tree lizard now known as *Urosaurus ornatus*, with *Uta stansburiana* or with *Urosaurus graciosus*—as I do until I have this diminutive,

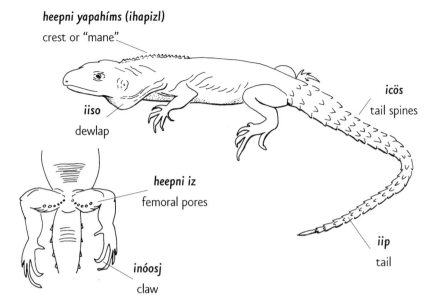

heepni yapahíms (ihapizl)
crest or "mane"

iiso
dewlap

heepni iz
femoral pores

inóosj
claw

icös
tail spines

iip
tail

Figure 14 (*left*). Spiny-tailed iguana parts named in Cmique Iitom.

Figure 15 (*below*). Chuckwalla body parts named in Cmique Iitom.

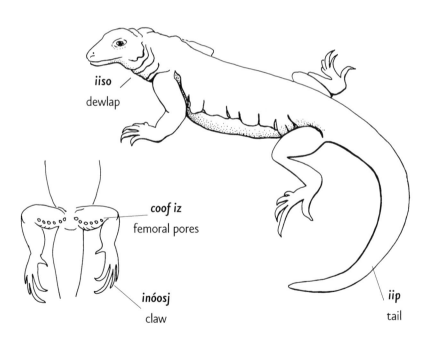

iiso
dewlap

coof iz
femoral pores

inóosj
claw

iip
tail

TABLE 13

Characteristics Used by Comcáac to Distinguish Sea Turtles

Species	English Common Name	Folk Taxon	Color	Claws per Limb
Caretta caretta	Loggerhead	xpeyo	orange/reddish-brown	one
		moosni ilítcoj caacöl	yellow, cream below	one
Chelonia mydas[1]	Green Sea Turtle	moosnáapa	gray-brown to gray-green	one
		moosníctoj	orange or pink albino	one
Eretmochelys imbricata	Hawksbill	moosni quipáacalc	red-brown, yellow below	one
		moosni sipoj	brown-black, white below	one
Lepidochelys olivacea	Pacific [Olive] Ridley	moosni otác	toad green, yellow below	one
Dermochelys coriacea	Leatherback	moosnípol	blue-black or charcoal green	none

[1]A number of Green Sea Turtle taxa are not included, for lack of sufficient detail recorded on their diagnostic traits.
[2]The weight of Green Sea Turtles varies greatly.

dull grayish lizard in hand. And yet they call Urosaurus ornatus by a distinctive name, hehe iti cooscl, "gray thing up in a tree." By using this habitat marker, the same one English-speaking herpetologists use, they distinguish it from other haquímet, "small lizards" such as tozípla, Side-blotched Lizards (Uta stansburiana).

In distinguishing one lizard or snake from another, the Comcáac use the animals' characteristic habitats and behaviors as much as they use morphological traits, a fact that should come as no surprise. Ethnobiologist Gene Hunn (1977) described a similar reliance on multiple diagnostic indicators among the Tzetzal Maya of Chiapas, Mexico, to tell different birds apart. Each bird, he learned, had a "characteristic range of shape and size, distinctive vocalizations, distinctive habits, and perhaps other characteristics of which I am not consciously aware."

Carapace Traits	Head Traits	Weight	Taste	Habitat and Groups
tapered toward back	large	80–150 kg	tasty	straits, alone or small groups
tapered toward back	smaller than *xpeyo*, green	100–200 kg?	less tasty than *xpeyo*	high seas to south
well-incised sections tightly fit to body	small	varies[2]	best tasting of all	straits, grouped rarely with Leatherback
well-incised sections	small	varies[2]	good	straits, grouped
thatched	thin beaky	80–130 kg	mild	straits, alone
thatched	thick beaky	80–130 kg	mild	straits, alone
flat convex, four front scutes flared	small	25–30 kg	inferior taste	high seas, in groups
seven ridges, pitched, rubbery, oval-shaped	large, blotched	500–800 kg	taboo to eat	high seas, rarely with green, alone or in twos

When I have asked certain knowledgeable Seri turtle hunters how they distinguish one kind of moosni from another, each has offered me a different mixture of criteria. For lack of space, I have "collapsed" or synthesized all the characteristics mentioned to me in one diagram (Table 13), even though no single turtle hunter offered all the indicators listed. Numerous characters come into play: body size and weight; carapace shape, color, and texture; plastron and flesh color; flipper size and claw number; head size and beak shape (Fig. 16); diet; seasonal habitat preference; sociability; docility; and the ability to understand Cmique Iitom. However, it is clear that a single trait, such as the overlapping plates on the carapace of the Hawksbill Turtle, can sometimes override other criteria to give an instant diagnosis.

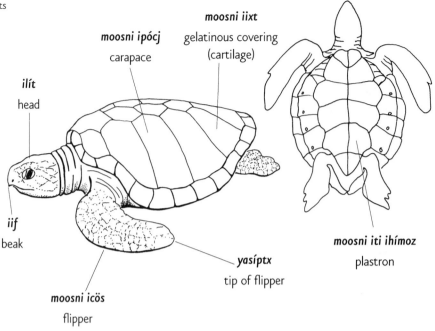

Figure 16. Sea turtle parts named in Cmique Iitom.

ilít
head

iif
beak

moosni ipócj
carapace

moosni iixt
gelatinous covering
(cartilage)

moosni icös
flipper

yasíptx
tip of flipper

moosni iti ihímoz
plastron

"The osprey-like turtle, it's almost the same as the turtle with overlapping plates, but its head is much more narrow. It lives in the bottom of the Canal [del Infiernillo], just as the Green Sea Turtle does—almost the same [in habit]. It has two claws on its flipper instead of one. The Green Sea Turtle seldom has more than one. It's never very big, maybe eighteen kilos on the average."

RAMÓN PERALES, *Punta Chueca*

Comcáac taxonomy identifies no less than eleven distinct taxa of sea turtles. For the Green Sea Turtle that Western scientists call *Chelonia mydas*, the Comcáac recognize eight morphs, four of which may be considered seasonal migratory phases rather than distinct populations or races as Felger and Moser (1985) suggested. The Comcáac recognize two races of Loggerheads and two races of Hawksbills as well. As Table 13 demonstrates, these populations or races are distinguishable from one another on multiple criteria, including morphological, behavioral, and ecological indicators. Compared to the current Western scientific taxonomy, the Comcáac taxonomy of sea turtles appears to "overclassify" them, discerning the equivalent of what domesticated-livestock geneticists might call races, stocks, or breeds. Ethnobiologists would call these named variants "folk varieties."

This differentiation of sea turtles into so many folk varieties would be unremarkable were it not for the fact that the tendency toward overclassification has otherwise been recorded only among horticulturists who select distinctive heirloom varieties of fruits and vegetables and among livestock herders who keep many breeds of cattle, sheep, fowl, or horses. It thus casts doubt on a long-held dogma that any culture which recognizes many taxa at the ranks of folk species and varieties will inevitably be sophisticated in animal husbandry or horticulture. However, Gene Hunn (pers. comm.) argues the opposite: perhaps Comcáac taxonomy is influenced by historic Seri participation in farming and/or ranching, or by their translocation of wild plants.

Just prior to completing his masterwork, *Ethnobiological Classification: Principles of Categorization of Plants and Animals in Traditional Societies* (1992), linguistic anthropologist Brent Berlin discussed Felger and Moser's (1985) writings on Seri plant classification with Hunn and felt obliged to insert one last section before his epilogue. He entitled it "The Seri: A Counterexample." Referring only to Seri plant classification and not to their detailed elaboration of sea turtle varieties, Berlin suggested that the Comcáac tendency to overclassify culturally valuable wild species

cast[s] serious doubt on [his theory's] validity as currently formulated. . . . In spite of their nonagricultural form of subsistence, the Seri exhibit a system of ethnobiological classification which conforms closely to that which one has come to expect in well-established horticultural populations. . . . On the basis of the Seri case, one is drawn to the less than satisfactory conclusion that the presence of well-established folk species does not unambiguously signal that the processes of domestication have been set in motion. While it still appears likely that specific taxa appear late in the evolution of systems of ethnobiological classification, and that domestication plays a major role in leading to their development, the Seri clearly manifest the same intense interest in looking closely at nature as do experimental horticulturists. . . . Nonetheless, the Seri remain a major ethnobiological puzzle that can be resolved only by the collection of new data, not only among this group but among other non-cultivating populations elsewhere.

It is curious to me that a scholar as eminent as Berlin considered the non-agricultural Comcáac unlikely to have an intense interest in intraspecific variation. By assuming that Comcáac taxonomy must fit the lower rungs of the evolutionary ladder of hunter-gatherers unable to achieve the more sophisticated biological classifications of farmers and herders, he inadvertently comes close to essentializing the Comcáac as social Darwinist W. J. McGee did a century ago. After a couple of weeks visiting with a few Seri at Rancho Costa Rica and observing their abandoned camps on Isla Tiburón, McGee (1898) wrote, "Certain characteristics of the [Seri] tribe strongly suggest a lowly condition, i.e., a condition approaching that of lower animals, especially of the carnivorous type."

Whether one counts the Comcáac as sophisticated taxonomists or as unrefined ones, living high on the evolutionary ladder or huddled down at its base, each of these biases presumes that all Seri individuals perceive, classify, and interact with their world's creatures in a uniform manner—a presumption that part 2 clearly dispels. Until recently, the heterogeneity of opinions, perceptual foci, and insights found within the Comcáac community have been swept under the rug (e.g., Romney, Weller, and Batchelder 1986; Rosenberg 1997). Or at best, it is assumed that only elders have the "right" answers to taxonomic questions, whereas younger residents, being more acculturated, are poorly versed in the logic that Hant Iha Quimx explained to their forebears. While it is to some extent true that younger Seri individuals frequently veer from the traditional consensus, Comcáac elders seldom reach unanimity, settling more often for provisional consensus.

What is delightful, albeit occasionally frustrating and ultimately human, about working in Comcáac communities is the heterogeneity of Seri perspectives, even on a matter as simple as naming a snake. As the species accounts in part 2 demonstrate, Seri individuals do not see the natural world with just one eye, or name its snakes, turtles, and lizards with just one mind and tongue.

reptiles as resources, curses, and cures

How the Comcáac Recognize Beauty, Utility, and Danger

Poisonous snakes have been an important source of mortality on almost all societies throughout human evolution. Close attention to them, enhanced by dream serpents and the symbols of culture, undoubtedly improves the chances of survival. . . . [As an] example of gene-culture coevolution, the frequency with which dream serpents and serpent symbols inhabit a culture is seen to be adjusted to the abundance of real poisonous snakes in the environment. But owing to the power of fear and fascination given them . . . they easily acquire additional mythic meaning: they serve in different cultures variously as healers, messengers, demons, and gods.

EDWARD O. WILSON, *Consilience: The Unity of Knowledge* (1998)

I had camped and bug-watched among the boojums in the Sierra Bacha off and on all summer long. On one of these trips, seven of us set up a makeshift kitchen under a palm-thatch *palapa* on the coast thirty kilometers north of Desemboque, well within the historic homelands of the Comcáac. Each evening at sunset, while we were out looking for nocturnal lizards and snakes, Sidewinders slithered into camp and curled up amid our sleeping bags and backpacks. When we returned, we caught them with our long-handled grippers and escorted them out of the kitchen, taking them several hundred meters down the volcanic bench to a stretch of sand just above the beach.

The night following this snake roundup, the very same Sidewinder—or perhaps one of its cohorts—would be back. They seemed to want the comforts that human company might offer, at least in the form of food scraps and cover, which we provided free of charge. Typically restricted to sandy washes, dunes, and beaches, these Sidewinders seemed unperturbed by the volcanic pebbles, desert pavement, talclike silt, and groundcloths that formed our floor beneath the *palapa*.

Later, when I told my Comcáac friends how surprised I was to see Side-winders away from large stretches of sand, they seemed unimpressed. *Ctaamjij*, they said, will venture into other habitats if food there is abundant. The isopods scurrying above high tide, the whiptails on the beach of volcanic cobble, and all of our cast-off morsels must have offered them such amenities.

On one particular full-moon night, after evicting the Sidewinders from our bedrolls, we went searching for night lizards. Our hosts, the managers of the Sierra Bacha game reserve, had recently bladed more than a hundred kilometers of new roads to ensure easy access to trophy hunters in pursuit of Desert Bighorn Sheep. Their bulldozers had knocked down dozens of ancient elephant trees, Organpipe, and Cardón cacti. These rotting trunks formed the perfect microhabitat for the Desert Night Lizard (*Xantusia vigilis*), a cryptic creature no larger than my index finger.

With flashlights, shovels, and prying rods in hand, we meandered our way through a glorious forest of towering Cardón and Organpipe, paying less attention, however, to these living giants than to their recently fallen kin. I turned over dozens of cactus carcasses before deciding to have a go at the rotting hulk of an elephant tree. As I pried it up, exposing the decomposing duff below, Rita Mehta, an intern from the University of California at Berkeley, squealed with delight: "We've got something here!" She pinned down a night lizard in the duff and quickly identified it as a *Xantusia*.

We all took a good look at it in the flashlight glow, then put it in an empty cooler. We would wait until the next day to release it, so that we could show it to Comcáac families during a brief visit to Desemboque.

When we arrived at the village, a half dozen of our Comcáac friends were sitting in the shade of an enormous salt cedar. After exchanging greetings and news, we showed them the rattlesnake roadkills and cast-off skin molts we had found since our previous visit. Neither the men nor the women seemed riled as I pulled dead snakes out of bags and coolers—

the same sleight-of-hand that had always sent my O'odham friends to the north shrieking and running for refuge. In contrast, the Comcáac talked among themselves about the sizes and colorations of the various snakes, offering me a brief summary in Spanish of what I could not understand in Cmique Iitom.

At last it came time to open the cooler where the night lizard lurked. I let the oldest Seri man present take a peek inside. *"Cozíxoj!"* he gasped. "Why did you pick that up? It's as dangerous for you as it is for us!" he whispered, backing away from the cooler. "Where did you get it?"

This confirmed what I had suspected: not only did night lizards and geckoes share the same name in Cmique Iitom, but the Comcáac bore the same fear for both nocturnal creatures.

"If a pregnant woman touches or even glances at this *cozíxoj*," one man warned, "the baby will grow up so skinny that its ribs will show." Someone else interjected: "It has a certain energy—one that affects even the unborn." Still another added: "You'll die if it sinks its teeth into you—its tongue is forked."

I thought of the scolding I'd received the first time I tried to show pickled reptile specimens to a Seri family in Punta Chueca. As I dumped a formaline-dowsed gecko out of a quart jar, Ernesto Molina slapped his hand over it before any of the women could see it: "No quiero ver ni tentar ese anima-lito," he muttered. Not only did he refuse to let anyone else lay eyes on it, but he wanted to deny its very existence. He then warned me that geckoes are especially dangerous for pregnant women to see, perhaps for the sake of one parturient woman who was among us. The translucent, fetuslike form of the reptile, he explained, could cause an expectant mother to envision the baby being born prematurely. Alternatively, the mother might have great difficulty in labor and delivery or, worse yet, have a stillborn.

The night lizard was not as embryonic-looking as a gecko, and there were no pregnant women among us at Desemboque, so I was not concerned that Ernesto's caution would apply here. But when those gathered began to tell about how it squirts a liquid that causes flesh to decay, I became hesitant to let anyone else see it. I put the cooler away, and in another hour or so the night lizard was back under its elephant tree trunk.

I was struck by the irony of it all. While the mere sight of certain animals would terrify my own aunts or my O'odham neighbors, the same sickening feeling came to my Comcáac friends not with rattlesnakes but with geckoes. Not with seasnakes but with night lizards. Not with coralsnakes but with Desert Iguanas. And not with Gila Monsters but with collared lizards.

To the Comcáac, it does not matter that the former creatures deliver toxic bites while the latter are known to be nonvenomous. Their fear of particular reptiles is rooted in the notion that considerable psychosomatic harm can come to anyone who does not behave appropriately around geckoes, night lizards, Desert Iguanas, collared lizards, and even their usually benign friends horned lizards, Desert Tortoises, and Leatherback Turtles.

Danger, like beauty and utility, is in the eye of the beholder.

TO PARAPHRASE pioneering cultural geographer Carl O. Sauer (1978), the choice to treat a creature as a useful natural resource or as a diabolical curse is ultimately a cultural assessment; it is not based merely on the attributes of the organism itself.

It is well known that many cultures around the world curse snakes and dread contact with them. Obviously, some ancient survival value is embedded in this stance. Yet why are the levels of abhorrence so extreme? New scientific interpretations of our dread of snakes—known technically as ophidiophobia—were recently generated as part of larger discussions regarding the so-called biophilia hypothesis (Kellert and Wilson 1993), the idea that humans have a genetically predisposed affinity for learning about and showing affection (or dread) for certain kinds of flowers and wildlife. E. O. Wilson (1998) interprets the typical human reaction of fear toward snakes as proof that biophilia could be innate:

Human beings [like chimpanzees] possess an innate aversion to snakes, and, as in the chimpanzee, it grows stronger during adolescence. The reaction is not a hard-wired instinct. . . . Children simply learn the fear of snakes more easily than they remain indifferent or learn affection for snakes. . . . [The] tendency to avoid snakes grows stronger with time . . . [but] the neural pathways of snake aversion have not been explored. We do not know the proximate cause of the phenomenon except to classify it as "prepared learning." In contrast, the probable ultimate cause, the survival value of the aversion, is well understood. . . . For hundreds of thousands of years, time enough for genetic changes in the brain to program the algorithms of prepared learning, poisonous snakes have been a significant source of injury and death to human beings. The response to the threat is not simply to avoid it, in the way that certain berries are recognized as poisonous through painful trial and error, but to feel the kind of apprehension and morbid fascination displayed in the presence of snakes by non-human primates.

In one study, lifelong residents above the Arctic Circle, far beyond the range of venomous snakes, suffered dramatic changes in blood pressure and heartbeat when a photo of a snake was flashed before their eyes. This nearly universal psychological reaction persisted longer and was more severe than the same subjects' responses to photos of guns, car accidents, and other life-threatening perils of the modern world. At the same time, we know that many zoologists, owners of pet reptiles, and medicine show hawkers known as *culebreros* have somehow overcome this innate dread of serpents.

Receiving little attention, however, is the fact that certain cultures seem to interact with venomous reptiles in relative comfort, and possess a rich serpentine symbology as well. This is not to say that all members of such a cultural community will handle reptiles or will refrain from expressions of fear under all conditions. On the contrary, a Western Diamondback Rattlesnake or Gila Monster hidden alongside a trail inspires fear in almost everyone. And yet, some people tend to regain their composure quickly after sighting a venomous reptile, and deal with the situation in a rational manner.

"Paaza—the Gila Monster—no, I don't think it has ever poisoned us. But it is muy cabrón—ornery—just like your friend Santiago [laughing]."
IGNACIO "NACHO" BARNET, *tribal governor, Punta Chueca*

It is remarkable how many men, women, and children in the Comcáac community handle venomous and nonvenomous snakes, dead or alive, without the flinching of their non-Seri neighbors. They have often had to evict rattlesnakes from their camps or kill them with stones or sticks to process the skins, meat, oil, and bones for practical uses. Seri fishermen like Roberto Camposano sometimes return to their villages with hundreds of snake carcasses obtained during camping trips to the islands. Certain Seri women speak with amusement and mock horror at their own prowess in hunting snakes, taking pride in the number of rattlers and gopher snakes they have killed and processed. Children grow up among elders who carry, skin, butcher, and roast or boil snakes, iguanas, and turtles without blinking an eye, and boys acquire these skills themselves at an early age (Figs. 17 and 18). They learn stories and songs about serpents, some of them mythical giants who live high in the mountains or deep within the ocean. They see rattlesnake designs elaborated in basketry and carvings (Fig. 19), and women wearing rings made from Gila Monster and Western Coralsnake skins. Clearly, one cannot consider the Comcáac as ophidiophobic, or snake-aversive, as many of the other cultures surrounding them.

"Sure, I get oil out of rattlesnakes [that I kill]. I cook the cocázni meat up just like you would do pork cracklings. You do it in water to render and flavor the oil, which you skim off the top with a spoon. Then, well, you clean up the bones from the same snake, for making necklaces. Yes, you can do the same thing with a Sidewinder. Yes, yes, I have."
MARÍA DEL CARMEN DÍAZ, *Desemboque*

Figure 17. A Comcáac boy with recently hunted Desert Tortoise and snake at a hunting camp near the center of Isla Tiburón. (Photo #7 by William N. Smith, 1946–51, courtesy of University of Arizona Special Collections)

Over the years, I have attempted to record Comcáac community members' opinions on which reptiles, including snakes, are perceived as dangerous, useful, or beautiful. Although these categories are not mutually exclusive, some students of cultural perceptions of wildlife make them out to be. By asking respondents only which animals they "like" or "dislike," or which they consider

Figure 18. Comcáac women preparing to butcher sea turtles following a successful hunt, with dogs present to eat the entrails, on the Canal del Infiernillo coast of Isla Tiburón. (Photo #15 by William N. Smith, summer/fall 1947, courtesy of University of Arizona Special Collections)

to be "good" or "bad," they set up dichotomies that fail to express the true complexity and ambiguity of people's attitudes about the natural world.

As Table 14 demonstrates, many of the reptile species that the Comcáac consider dangerous are also valued for their beauty or utility. In assessing Comcáac perceptions of reptiles, I considered several indicators for each of these quali-

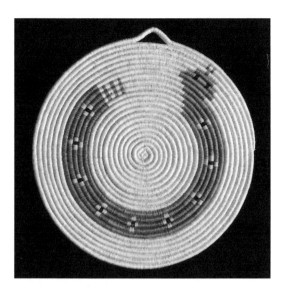

Figure 19. A rattlesnake design in a woven plaque. (Photo Helga Teiwes, 1999)

ties, not only what individuals said, but also what they expressed through their actions, art forms, and taboos.

Only rarely among the Comcáac does recognition of danger preclude a reptile being valued for its other qualities. In fact, according to the ethnozoological survey conducted by Borys Malkin (1962) in the 1950s, the percentage of reptile species used by the Comcáac in the region was as high as that for any other comparable "life-form": marine invertebrates, fish, terrestrial insects, amphibians, birds, or mammals.

If ophidiophobia does occur among the Comcáac, it appears to be overridden by the utility and beauty of local reptiles—with one exception: the Desert Iguana is considered so dangerous that it is not used in any way or represented in any traditional art form (other than in a "horror story"). The few Seri individuals who would even talk to me about Desert Iguanas expressed far less familiarity with them in terms of behavior and morphology than with other iguanids. The Comcáac are shocked when they hear that their Hia c-eḍ O'odham neighbors to the north historically ate Desert Iguanas. When asked why they fear the animal so, they tell a story in which a Desert Iguana is the suspected assassin of a Seri wayfarer traveling between two camps. Few people

TABLE 14

Reptiles Associated with Beauty, Utility, and Danger

Species	Comcáac Name	Associated with Beauty[1]	Associated with Utility[2]	Associated with Danger[3]
TURTLES AND TORTOISES				
TESTUDINIDAE				
Gopherus agassizii	xtamóosni, ziix hehet cöquiij, ziix catotim	•	•	
KINOSTERNIDAE				
Kinosternon sonoriense	xtamáaija	•		
CHELONIIDAE				
Caretta caretta	xpeyo, ilítcoj caacöl	•	•	
Chelonia mydas	moosni, ziix xepe ano quiih	•	•	
Eretmochelys imbricata	moosni quipáacalc, moosni sipoj	•	•	
Lepidochelys olivacea	moosni otác	•	•	
DERMOCHELYIDAE				
Dermochelys coriacea	moosnípol, xica cmotómanoj	•	•	
LIZARDS				
CROTAPHYTIDAE				
Crotaphytus dickersonae	hast coof			•
C. nebrius	hast coof			•
IGUANIDAE				
Ctenosaura conspicuosa	heepni	•	•	
Dipsosaurus dorsalis	meyo, ziix tocázni heme imocómjc	•		•
Sauromalus hispidus	coof coopol	•	•	
S. obesus	coof, ziix hast iizx ano coom	•	•	
S. varius	coof	•		
PHRYNOSOMATIDAE				
Callisaurus draconoides	ctamófi, ctoicj, matpayóo	•		
Phrynosoma solare	hant coáaxoj	•		•
Sceloporus magister	haasj	•		
Urosaurus ornatus	hehe iti cooscl	•		
Uta stansburiana	yax quiip, tozípla	•		
EUBLEPHARIDAE				
Coleonyx variegatus	cozíxoj	•		
GEKKONIDAE				
Phyllodactylus xanti	cozíxoj			•

TABLE 14 (continued)

Species	Comcáac Name	Associated with Beauty[1]	Associated with Utility[2]	Associated with Danger[3]
TEIIDAE				
Cnemidophorus tigris	ctoixa, ctoixa iipíil (hatchlings)	•		
XANTUSIIDAE				
Xantusia vigilis	cozíxoj			•
HELODERMATIDAE				
Heloderma suspectum	paaza	•	•	?
SNAKES				
BOIDAE				
Boa constrictor	xasáacoj	•		•
Charina trivirgata	ziix haas ano cocázni	•	•	
COLUBRIDAE				
Chilomeniscus stramineus	hapéquet camízj	•	•	
Hypsiglena torquata	coimaj coospoj			•
Masticophis bilineatus	coimaj, maxáa		•	•
Pituophis melanoleucus	cocaznáacöl	•	•	
ELAPIDAE				
Micruroides euryxanthus	coftj	•		?
HYDROPHIIDAE				
Pelamis platurus	xepe ano cocázni			•
VIPERIDAE				
Crotalus atrox	cocázni	•	•	•
C. cerastes	(cocázni) ctaamjij	•	•	•
CROCODILES				
CROCODYLIDAE				
Crocodylus acutus	xepe ano heepni, xepe ano paaza	•		?

[1]As defined here, includes evocation in songs, elaboration in drawings, carvings, basketry, or other visual arts, and perceived "loveliness" expressed in interviews.

[2]Includes use as food, ornament, utensil, medicine, or sellable item.

[3]Includes expressed disgust or fear when shown a live animal, photo, or drawing, as well as oral accounts of envenomation, biting, or taboos that if broken generate harm to the violator.

now refer to the Desert Iguana by its original name, *meyo*, but instead describe it as *ziix tocázni heme imocómjc*, "thing that, if it bites, prevents arriving at home camp." Some believe that if it bites you, it kills you on the spot.

"*Ziix tocázni heme imocómjc*—it lives in the *monte* [scrublands]—poisonous. Whoever is bitten by it won't ever come home again."
AMALIA ASTORGA, *Desemboque*

The capture or use of Desert Iguanas is simply not an option for the Comcáac. In general, it appears that some of the taboos associated with Desert Iguanas, Regal Horned Lizards, Leatherback Turtles, and Desert Tortoises are associated with ancient stories referring to the time when animals could understand Cmique Iitom and could themselves speak and argue. Seri individuals who fail to heed the message of these stories put themselves at risk, becoming susceptible to illness or even death. Just as their O'odham neighbors suffer from what they call "staying sicknesses" (Bahr et al. 1979), Seri who break a taboo and realize it often experience great anxiety and psychosomatic trauma that can be relieved only through a shamanistic purification ritual that restores healthy relationships with particular animals.

"Horned lizards have a way of protecting themselves. Should you disturb a horned lizard, it will shoot this little thing into you . . . like when a little spider strikes you at your knee or shoulder or elbow joint . . . but then this substance enters your body through that point, causing you great pain. Only a medicine man can clean your body of such a disturbance. You can only get well again through a curing ceremony."
ERNESTO MOLINA, *Punta Chueca*

The Comcáac do not necessarily claim that taboo-related illnesses are more serious than being bitten by a venomous animal, yet some Seri individuals have considerable tolerance for the presence of Gila Monsters and Western Coralsnakes, fearing them only under certain circumstances. Men will casually han-

dle coralsnakes and consider them docile. Indeed, few Western Coralsnake en-
venomations have been recorded in American deserts, and there have been no
known human deaths; while their venom is potent, the quantity of venom per
coralsnake is minuscule compared to that of rattlesnakes.

Most Seri individuals can readily distinguish true *coralillos* from false coralsnakes, having memorized the tongue twister mentioned in chapter 4 about
their color banding patterns. Nevertheless, these identifications are undertaken
with a certain nonchalance, and today young men are likely to call any of the
nonvenomous "false" coralsnakes *coralillos* without any expression of fear.

The Comcáac also recognize that Gila Monsters and Yellow-bellied Seasnakes
will attack humans, but only if they are trapped or molested. Maurice Richardson (1972) tells of a sailor swimming among a shoal of seasnakes without suffering any harm. The Seri fishermen I know who distinguish the Yellow-bellied
Seasnake from more commonly seen eels worry about it only when it is entangled in their nets, and justifiably so: more than 50 percent of the humans
known to have been bitten by seasnakes were attempting to extricate them from
fishing nets. Although the survival rate of recently hospitalized victims even
without treatment by antivenin is 92 percent (Thomas and Scott 1997), the
venom is extremely toxic—so much so that a fully envenomized bite may kill
a person within two and a half hours. Just four drops of pelagic seasnake venom
is enough to kill three men (Thomas and Scott 1997). The Comcáac may not
know these facts, but they do know to be watchful when hauling in nets. The
Comcáac sense of danger is not based on the mere presence of a venomous reptile, but on the context in which it is found.

"*Xepe ano cocázni* [Yellow-bellied Seasnake]—very dangerous, very poisonous. They'll
come up onto the beach in groups. Or they'll be in the shallows, in the sand. Bite
you on the feet, then you'll swell up. My brother-in-law was badly bitten by one
when he was diving for octopus. He survived, but his arm was swollen up for a week.
[It] really hurt."

MARÍA FÉLIX, *Desemboque*

Even despite precautions, Seri are occasionally the victims of reptiles that bite and envenomate. One time in Punta Chueca I spoke at length with Comcáac elder Miguel Barnet about his experiences with two envenomated snakebites. As we sat together in the sand under a ramada, Miguel rolled up his pants leg and showed me where he had been bitten in the calf. He spoke candidly of how for two months the bite remained open and discolored, exuding a watery liquid, despite the diligent care given him by family members.

During the course of our conversation two of the other seven people present also showed me where they had been bitten by Sidewinders or other rattlesnakes, and a third explained that she personally had treated several snakebite victims. In each case, the symptoms described were consistent with venomous bites: there was local deterioration of the flesh around the bite that took weeks to heal, as well as systemic effects such as nausea and the swelling of glands immediately following each incident. If I am correct in assuming that none of these were dry (non-envenomated) bites, it is all the more remarkable that the victims survived with only traditional treatments.

When bites occur among the Comcáac, victims are typically treated by family members or friends who are present at the time. "Medicine men" are not the only ones who have to know how to treat reptile bites, for often these calamities occur away from villages or encampments, where such specialists are unavailable.

It is my impression that venomous reptile bites occur far less frequently in Comcáac villages than stingray or scorpion stings or spider bites. Whatever the culprit, all stings and bites, venomous or otherwise, are today treated by many villagers with the same herbal treatments. Cutting, sucking, or tourniquets may also be used for snakebites, but not for other bites, punctures, or stings. Only a few elderly individuals specifically mentioned that some herbs were more effective for ray stings than for rattlesnake bites, or vice versa.

At least one historic visitor to Comcáac camps was appreciative of the folk treatments he received there. Three centuries ago, Padre Adamo Gilg was stung by a scorpion in the presence of Comcáac families that he was attempting to convert to Christianity. They apparently treated his wound by first burning it, then applying herbal plasters. He was so impressed by the efficacy of the remedy that in his 1692 testimony of his involvement with the Comcáac he cited their medical knowledge as one of many laudatory qualities: "The Creator of all things has revealed to them certain medicines, with which they easily heal themselves of illnesses, and especially of so many poisonous animals; as I myself experienced" (DiPeso and Matson 1965).

Table 15 inventories the plants the Comcáac consider to be useful for treating venomous reptile bites and their secondary symptoms. Some of the same plants that are used for treating bites by nonvenomous snakes such as kingsnakes are also used for bites suffered by horses and cattle. A flexible treatment strategy for all snakebites is not surprising, since toxicologists now recognize a continuum of risk from highly venomous snakes to nonvenomous snakes. As medical herpetologist Steve Grenard (1994) has demonstrated, some snakes formerly considered nonvenomous are now known to have mildly toxic bites; moreover, the Western medical differentiation between snakes with neurotoxins and those with hemotoxins is now considered to be a continuum, not a dichotomy. In addition, many reptile bites, whether envenomated or not, put the victim at risk for infection by microbes such as *Escherichia coli* and *Pseudomonas aeruginosa*, as well as from shock, local tissue injury, dramatic changes in blood pressure, and temporary neuromuscular damage (Grenard 1994; Thomas and Scott 1997).

When medical herpetologists have analyzed the effects of some of the seven hundred or so species of flowering plants used to treat snakebite, they have identified two patterns of influence (Ferreira et al. 1992; Houghton and Osibogen 1993; Johnson Gordon, Moreno Salazar, and López Estudillo 1994). First, as toxicologist Peter Houghton (1994) has observed, "the reputed activity of some of the plants might be explained, at least in part, by biological activities of relevance to the symptoms of snakebite rather than direct antagonism to the venom." Plants that have tranquilizing effects may reduce the symptoms of shock, while analgesic, anti-inflammatory, and antimicrobial properties may also serve to improve the overall status of the victim. Second, a few traditional medicinal plants (such as *Curcuma longa* in the family Zingaberaceae) have been found by Brazilian researchers Ferreira et al. (1992) to neutralize the hemorrhagic actions of venoms from rattlers and other snakes. More generally, tan-

TABLE 15

Plants Used in Curing Reptile Bites

Species	Comcáac Name	Spanish Name	Source	Comments on Treatment
ASTERACEAE				
Hymenoclea monogyra	caasol coozlil, caasol itac cöihíipe, caasol itac coosotoj	jécota, yerba del pasmo	Felger and Moser 1985	leaves boiled as tea to reduce swelling
H. salsola	caasol cacat, caasol coozlil, caasol zíix iic cöihíipe	jécota, yerba del pasmo	Felger and Moser 1985	leaves boiled as tea to reduce swelling
Palafoxia arida	moosni iha, moosni oohit	palafoxia	Felger and Moser 1985	leaves ground, poultice on bite
BURSERACEAE				
Bursera laxiflora	xoop caacöl	torote prieto	Nabhan and Monti 2000	bark used in mixture with *Larrea tridentata* as tea to calm pain
CACTACEAE				
Lophocereus schottii	hehe is quizil	senita, sina	Felger and Moser 1985; Nabhan and Monti 1997–99	stem slice used as poultice to absorb venom
CAPPARIDACEAE				
Wislizenia refracta	hehe iyapxot cmasöl	guaco	Nabhan and Monti 1997–99; Gentry 1938	leaves made into a tea to relieve pain
CHENOPODIACEAE				
Atriplex barclayana	spitj	yerba de mantarraya	Reina-Guerrero and Morales 1998	leaves and stems boiled and applied as wash to bathe bite
LAMIACEAE				
Lippia graveolens	xomcahíift	orégano	Nabhan and Monti 2000	leaves made into a poultice and wash to detoxify venom
PHYTOLACCACEAE				
Stegnosperma halimifolium	xneejamsíictoj	chapacolor	Felger and Moser 1985; Nabhan and Monti 1997–99	leaves chewed, poultice on bite, or as tea
RHIZOPHORACEAE				
Rhizophora mangle	pnaacoj-xnazolcam, pnazolcam, mojépe camoz (fruit)	mangle rojo	Felger and Moser 1985; Reina-Guerrero and Morales 1998; Nabhan and Monti 1997–99	fruit made into a tea to stop nausea
ZYGOPHYLLACEAE				
Larrea tridentata	haaxat	jediondilla	Felger and Moser 1985; Nabhan and Monti 1997–99	leaves charred; poultice to reduce swelling, or boiled as wash for bite

Note: Use rights reserved by the Comcáac community.

nins and other compounds of low molecular weight may react with peptides in venoms, likewise neutralizing some of their effects.

No clinical or laboratory evidence yet exists to show that the particular herbs or treatments used by the Comcáac function to neutralize nerve, heart, or muscle toxins associated with various reptile bites. By looking at the other medical problems for which the Comcáac use these same herbs, it is plausible that these remedies reduce the severity of swelling, nausea, bleeding, or secondary infection associated with bites. Some of the herbs used by the Comcáac are also used by other indigenous communities to treat the symptoms of swelling, localized pain, and nausea (Reina-Guerrero and Morales 1997; Johnson Gordon, Moreno Salazar, and López Estudillo 1994). Laboratory analysis has not yet been undertaken to assess their effectiveness. It may be that some of these herbs offer protection against secondary microbial infections potentially resulting from any kind of animal bite or sting.

One herb frequently used in snakebite treatment today by both the Comcáac and the villagers of San Joaquin, Baja California, is not mentioned in Felger and Moser's "Seri Indian Pharmacopoeia" (1974) or in other medicinal plant inventories from indigenous Sonoran communities (Johnson Gordon, Moreno Salazar, and López Estudillo 1994): *Wislizenia refracta* var. *palmeri*, known as *guaco* in Spanish and informally described as *hehe yapxöt cmasol*, "plant with yellow flowers," in Cmique Iitom. More than sixty years ago, my mentor Howard Scott Gentry (1938) recorded that in San Joaquin, *guaco* was "reputed to have medicinal properties; herbage concocted and drunk or used as a wash for snake bites." While it is possible that the similarity in use reveals an ancient relationship between the Comcáac and Baja California peoples, more likely it simply means that the Comcáac have traded remedies with the many Baja California fishermen who have worked on the Sonoran coast and Gulf islands during the last fifty years. If the descriptive name is any indication, this plant is such a recent arrival to Seriland that it is simply described as one of many yellow-flowered plants in Cmique Iitom.

Table 16 highlights stages of traditional diagnosis and treatment, such as the use of teas, tourniquets, pressure bandages, poultices made of leafy herbs or of columnar cactus stem sections, and the sucking of venom. Notably, the same

TABLE 16

Stages in Reptile Bite Diagnosis and Treatment among Contemporary Comcáac

Triggering Condition	Responses	Comments
A reptile bite occurs	Remove victim from presence of reptile and attempt to identify (or kill) it	Because a bite can occur anywhere, when one is alone or in a group, the capacity to capture, kill, or even identify the reptile is situation-specific.
Determination made that reptile envenomated the victim	1. Suck out venom as quickly as possible (unless the care provider's gums or teeth are infected) 2. Use tourniquet to reduce flow of venom into body 3. Apply "antivenin herbal compact" or slice of senita cactus on wound 4. Take victim to hospital or clinic in nearby town to obtain antivenin serum if available	Sucking venom and using tourniquets are infrequently done today, especially if a hospital is within forty minutes' reach; however, the travel time is often much longer and hospitals seldom have antivenin serums in stock, so herbs are often applied in the field or in transit. The herb choices are typically those closest at hand; others may be preferable but less accessible because they are found only in a few localities away from the villages. Senita (*Lophocereus schottii*) is applied to reduce pain and to absorb venom from a cross-cut bite. Anyone may administer.
Determination made that reptile did not envenomate or was nonvenomous	Concentrate on reducing bleeding and secondary symptoms	Some of the same herbs may be used for treating secondary symptoms of both venomous and nonvenomous bites and stings.
Profuse bleeding and swelling occurs	1. A bandage is held on wound and changed frequently 2. Herbs are boiled in two liters of hot water, then rags soaked in the brew are placed on the wound and changed every five minutes 3. The victim is put into resting position	A wider variety of plants known to help swelling and blood coagulation may be mixed into the brew, depending on their immediate local and seasonal availability. Some women keep dried, ground leaves and flowers (but not roots) for such emergencies. This step need not be administered by a *curandera*; any family member or co-worker available may need to initiate these treatments.

herb is often drunk as a tea and applied as a wash or poultice to the wound. Since the 1970s, these methods have been used by the Comcáac more as stopgap measures until other medical assistance can be obtained from hospitals and emergency first aid clinics an hour or so from Comcáac villages. Ironically, however, these institutions seldom keep laboratory-produced antivenin serums on hand, rendering "stopgap" measures of critical importance. In many ways, the Comcáac predicament is much the same as that of the Awa communities of coastal Colombia, whose dilemma is so elegantly summarized in the ethnomedical monograph *Por el camino culebrero* (Parra-Rizo and Virsano-Bellow 1994): "To conclude, it is worth noting that neither [the local health center] nor [the hospital in the nearest town] are supplied with antivenin serums. The herbal antidotes and wound washes that are prepared by the Indians and by some mestizos have historically been and remain today—after some twenty thousand years of human presence in the Americas—the only resources that the people of this rural region possess to counteract snakebites."

Most Western medical practitioners urge that snakebite victims be spared the time required to administer folk remedies so that all efforts are focused on transporting the victims to the closest source of antivenin as soon as possible. However, one documented account from a remote Latin American community confirms that a snakebite victim's condition can be stabilized much more quickly when indigenous curing practices are combined with Western emergency medical care (Zethelius and Balick 1982). As long as herbal treatments do work to reduce the swelling, bleeding, nausea, or anxiety associated with reptile bites (regardless of whether they counteract the venom itself), they play a critical role in elevating a victim's chances of survival. In the case of Comcáac fishermen, the mere fact that they spend time in remote island habitats where snakes are abundant—and hospitals may be many hours away—justifies training fishermen in traditional herbal as well as conventional emergency medical treatment of snakebites.

For these reasons, the Comcáac (along with other indigenous peoples) should be guaranteed the intellectual property rights associated with any future pharmacological "discoveries" derived from their knowledge of the herbal treatments used for reptile bites, which may well prove applicable to other medical

Figure 20. Rattlesnake oil still is found in products sold in Sonoran marketplaces. (Photo David Burckhalter, 1999)

emergencies as well. A Comcáac collective of women who gather and package herbs is currently exploring and articulating ways in which their intellectual property rights are legally protected by the Global Convention on Biodiversity. While they have chosen to offer herbal packages to non-Indians, the packages are labeled in a manner that asserts their rights to exclusively gather, prepare, and market their traditional herbal remedies within the boundaries of the Republic of Mexico. I encourage all readers to respect these rights, which this book clearly documents as belonging to the Comcáac.

It may be easy for physicians safely ensconced in their offices and near clinics harboring quantities of antivenin to dismiss indigenous remedies for snakebite. It's even easier for them to criticize the medicinal use of snake oil, a substance still employed by the Comcáac (Fig. 20). Both popularly and among professionals in the West, the very term *snake oil* is nowadays synonymous with quackery among folk healers (Fowler 1997), and as a consequence, most contemporary enthnomedical researchers have shied away from attempts to evaluate its efficacy. The history of public perceptions of snake oil remedies in the Americas is an interesting one, revealing both the romanticism and the cynicism with which European-Americans have regarded indigenous cures. In that sense, it has direct bearing on how the Comcáac perceive their relationships to reptiles and their products, and so warrants brief examination here.

"I make a powder from the dried rattlesnake meat. The skins are worth a lot at a fair I go to, where I also sell a pomade of rattlesnake oil in four-ounce Gerber baby food jars. It cures sore throats and earaches. I learned how to make it from some Tarahumara people. They say that even the venom is good medicinally. It serves to cure cancer."

AMALIA ASTORGA, *Desemboque*

In the decades immediately following the Treaty of Guadalupe Hidalgo (1848) and the Gadsden Purchase (1853), the deserts of the U.S.–Mexico borderlands were overrun with cross-cultural hawkers who sold snake oil through their medicine shows, extolling it as a cure for innumerable ills. During the World Columbian Exposition of 1893 in Chicago, one such huckster, Clark "The Snake Oil King" Stanley, who was an exposition celebrity, killed and processed hundreds of rattlesnakes, claiming to be rescuing ancient medicinal wisdom to serve the masses newly arrived from metropolitan areas. Stanley testified that he had learned a special sacred formula for his snake oil liniment—which he stated was effective against rheumatism, colds, and general aches and pains—while reputedly living among the Hopi in 1879. However, when federal chemists seized a shipment of his concoction in 1917 to analyze its contents, they found it to be a light mineral oil doctored with 1 percent beef fat, red pepper, and traces of camphor and turpentine (Fowler 1997). The news of this scandal fostered cynicism among the literate public of North America, leaving them skeptical of any folk healers swearing to the efficacy of snake oil and kindred products.

Despite such cynicism, which is now shared by some younger members of Comcáac communities, traditional medical practitioners continue to use rattlesnake oil to treat a variety of maladies. As one indigenous healer said to ethnobiologist Mark Plotkin (2000), "What's all this fuss about snake oil? I'll use it as long as it works!"

It came as some surprise to me to learn that the Comcáac medicinally use more species of reptiles (15) than they use species in all other orders of animals combined (9). Interpreted another way, twelve folk taxa of fauna are recognized by the Comcáac as having curative value, of which five are reptiles: *ziix catotim* (Desert Tortoises); *moosni* (sea turtles); *cocázni* (rattlesnakes); *coof* (chuckwallas); and *paaza* (Gila Monsters) (Table 17).

Felger and Moser's "Seri Indian Pharmacopoeia" (1974) was the first to document in print several of these uses of reptiles. A more recent ethnomedical survey by Sonoran ethnobiologist Ana-Lilia Reina-Guerrero and Comcáac folklorist David Morales (1997) identified still other uses of reptiles, describing them both in Spanish and in Cmique Iitom.

The intensity of medicinal uses of reptiles among the Comcáac is unparalleled anywhere in northern Mexico or the U.S. Southwest, but reptiles enter into Comcáac material culture in other ways as well. Reptiles such as sea turtles, tortoises, iguanas, and chuckwallas are also important to the health of the Comcáac as nutritional resources. Over a century ago, ethnographer W. J. McGee (1898) contended that reptiles formed the core of Seri subsistence, and that sea turtle meat alone accounted for a quarter of all food consumed by the coastal Seri.

"The tribesman and the turtle have entered into an inimical communality something like that of Siouxan Indians and buffalo in olden time, whereby both may benefit and whereby the more intelligent communal certainly profits greatly. The flesh of the turtle yields food; some of its bones yield implements; its carapace yields a house covering, a convenient substitute for umbrella or dog-tent, a temporary buckler, and an emergency tray or cistern as well as a comfortable cradle at the beginning of life and the conventional coffin at its end; while the only native footgear known [sic] is a sandal made from the integument of a turtle flipper."

W. J. MCGEE, *The Seri Indians* (1898)

When certain Comcáac elders speak of eating the meat and drinking the blood of these reptiles, they convey a sense that these foods are esteemed in a manner that approaches being sacramental. This is perhaps the reason that turtle,

TABLE 17

Reptiles Used in Curing among Contemporary Comcáac

Species	Comcáac Name	Illness or Injury	Parts Used	Source
TURTLES AND TORTOISES				
TESTUDINIDAE				
Gopherus agassizii	xtamóosni, ziix hehet cöquiij, ziix catotim	anemia	meat	Reina-Guerrero and Morales 1998
CHELONIIDAE				
Chelonia mydas	moosni	craziness, skin rash, sore throat, infertility, sea and car sickness	penis, intestine, blood, fat (of young migrant)	Reina-Guerrero and Morales 1998; Felger and Moser 1985
LIZARDS				
IGUANIDAE				
Sauromalus spp. (incl. S. obesus)	coof (ziix hast iizx ano coom)	skin rash	fat	Reina-Guerrero and Morales 1998
HELODERMATIDAE				
Heloderma suspectum	paaza	headache	skin	Reina-Guerrero and Morales 1998; Felger and Moser 1985
SNAKES				
VIPERIDAE				
Crotalus spp. (incl. C. cerastes)	cocázni (ctaamjij)	skin allergy, sores, colds, tuberculosis	fat, skin, rattle, meat	Reina-Guerrero and Morales 1998; Felger and Moser 1985

Note: Use rights reserved by the Comcáac community.

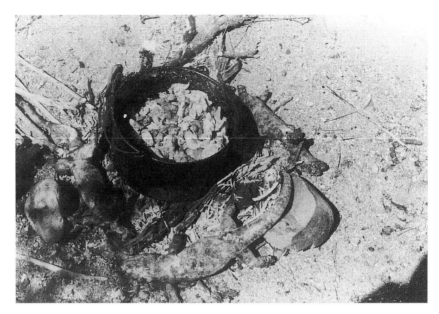

Figure 21. Green Sea Turtle meat and chuckwallas roasting, Isla San Esteban. (Photo #21 by William N. Smith, Sept. 20–21, 1947, courtesy of University of Arizona Special Collections)

tortoise, and iguana meat continue to be preferred over that of domestic live-stock and fowl (Figs. 21 and 22). Only the eating of Gila Monster tails has fallen completely out of Comcáac practice.

"Heepni? You look for those iguanas where they reside in the ironwood canopies, maybe in hollow branches there. You find them close to shore, in the *monte*, the thick desert scrub. Once you get one, you eat it all. Every last bit is tasty, as good as rattlesnake. Yes, I've personally eaten *heepni*."

ANTONIO LÓPEZ, *Desemboque*

McGee was not the only observer to rank sea turtles among the most im-portant components of the Comcáac diet. In the early 1950s, Borys Malkin (1962) placed sea turtles and other reptiles just below fish in their dietary importance. Richard Felger and Becky Moser (1985) also considered these animal foods main-stays of Comcáac nutrition, following mesquite and other legumes, cactus fruit, and fish. Today, global declines in sea turtle populations have lessened their nutritional importance in Comcáac communities, but Green Sea Turtles are still consumed ceremonially, or eaten if found injured in fishermen's nets.

The intensity of consumption of reptile foods by the Comcáac has always varied widely over space and time. Inland, at Pozo Coyote, Desert Tortoise was a mainstay, and the Sonoran Mud Turtle was eaten as recently as a half century ago. Isla San Esteban residents, in contrast, were said to be expert at hunting iguanas and sea turtles, though they somewhat disliked the taste of sea turtle, frequently eating it only during periods of food scarcity (Bowen 2000). According to what Seri elder Jesús Morales told archeologist Tom Bowen (2000), the taste of Black Chuckwalla was attractive enough to lure the San Esteban people over to the San Lorenzo islands now and then, despite the perils of this journey across the high seas.

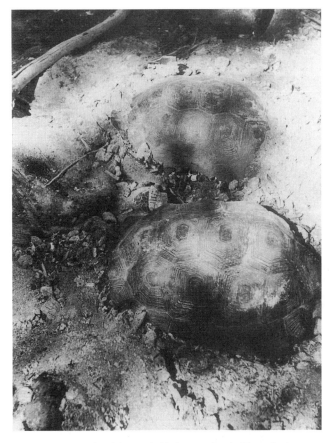

Figure 22. Two tortoises being cooked in ironwood coals. (Photo #24 by William N. Smith, n.d., courtesy of University of Arizona Special Collections)

"There are probably four men left who have regularly eaten chuckwallas over their entire lifetimes: Miguel Barnet, Jesús Martínez, José Manuel Romero, and José Luis Blanco. These days, the boys growing up here in the villages know only the shorelines of the islands, they never go into the interior to go hunting. Very few of them have even gone [hunting there]."

JESÚS ROJO, *Punta Chueca*

Figure 23. Sea turtle bladder and stomach. These traditionally provide storage for edible liquids and Green Sea Turtle oil (rendered from the fat at coastal camps on Isla Tiburón). The long container is the stomach of the Green Sea Turtle; the smaller is the bladder. (Photo #22 by William N. Smith, Oct. 1947, courtesy of University of Arizona Special Collections)

Desert Tortoise eggs contributed significant protein to the Comcáac diet seasonally, especially for those who historically lived inland from the coast, both on Isla Tiburón and on the mainland. Desert Tortoise eggs were commonly sought out until fifty or so years ago; even today, though less often gathered, they remain well regarded. The frequency of use of sea turtle eggs is more difficult to determine, for these were not harvested from beaches but were invariably "rescued" from harpooned turtles that turned out to be gravid. Younger Seri individuals now maintain that the consumption of turtle and tortoise eggs is taboo, even though elders in their families admit that they have eaten them on occasion.

Once a sea turtle or tortoise was killed, butchered, and its eggs extracted, other parts were utilized in a variety of ways. Turtle carapaces were stood upright to serve as windbreaks next to campfires, and were also used as sleds, trays, cradles, and as "thatch" on houses. Tortoise carapaces were used as rattles for ceremonial dances or carved to make rings. The bladders of certain sea turtles were used as watertight containers to store and carry liquids and small seeds (Fig. 23).

"The bladder which we originally used was that of *moosniscl*, the dark-shelled form
of the Green Sea Turtle. It was used for holding turtle oil, seeds, and water. . . .
No, it didn't spoil the liquid with its aroma. That's not true for the bladders of *xpeyo*,
the Loggerhead. It has its own peculiar smell and consistency that we didn't like."
ALFREDO LÓPEZ, *Punta Chueca*

Sea turtle bones were fashioned into a variety of objects, including dolls, rings, and figurines, some of which had sacred significance. According to William "Seri Bill" Smith (1951), both sea turtle bones and clay figurines shaped like them, excavated at various sites within Seriland, were once important in Com-cáac ceremonies relating to their creation story. Sea turtle bone dolls continue to be made today and are occasionally offered to tourists for purchase as a traditional craft item.

While "Seri arts and crafts" are now sold from Alaska to Mexico City, outsiders usually know Comcáac carvings and basketry only; they are seldom aware of other artistic expressions, both fleeting and enduring, such as bone dolls, beach drawings, face paintings, sand etchings, pictographs, or clay, stone, and elephant tree figurines. Each of these art forms includes representations of reptiles. Even the rattles of Sidewinders and other rattlesnakes find their way into "Seri jewelry" today, as much for locals as for tourists (Fig. 24).

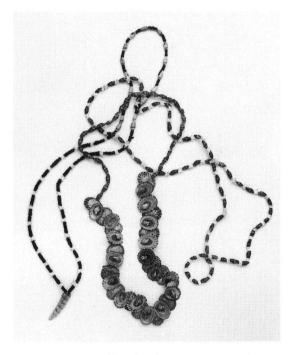

Figure 24. Comcáac necklaces for sale to tourists in Punta Chueca. Note rattlesnake-rattle "medallion." (Photo Helga Teiwes, 1997)

Images of sea turtles, for example, are often roughed out in the sand of beaches with a shell or rock stylus. When teenager Raúl Molina stumbled upon an ancient sandstone carving of a Leatherback Turtle partially buried in the sand near *Hast Moosni*, "Sea Turtle Rock," in 1992, he couldn't sleep at all that night. His father, Roberto Molina, was sure he had encountered an effigy that still retained much of its power.

In contrast to that sandstone Leatherback, which has survived the centuries, the geometric designs representing Leatherbacks that are painted on women's and girls' faces are intended to last for no more than the four days of the *xica cmotómanoj* ceremony (see chapter 7). According to Sonoran folklorist Alejandrina Espinosa-Reyna (1997), who lived with Comcáac families between 1984 and 1990, vegetable and mineral dyes are used in these designs to represent certain qualities and conditions. A red mineral ochre, *xpaahjö*, and the red sap from *xoop*, the elephant tree (*Bursera microphylla*), represent death. A white gypsum, *hantíxp*, attracts good luck. To represent the sea, a blue vegetable and mineral dye is mixed from the gum of *mocni*, the *guayacán* tree (*Guaiacum coulteri*); from *tincl*, the root of canyon ragweed (*Ambrosia ambrosioides*), or from *xcoctz*, the root bark of bursage (*Ambrosia dumosa*); and from a local stone called *iqui icóizj*. Those who arrive near where the ceremony is taking place and who don't want their faces to be painted with these Leatherback designs must leave, lest they bring bad luck to the entire group of celebrants.

"The Leatherback understands the language of the Seri, who sing this song with faces painted white—the white coming from a greenish rock—while we are waiting for the turtle to arrive [on the beach]."

JESÚS ROJO, *Punta Chueca*

In comparison with ancient sandstone carvings and ceremonial face paintings, the development of ironwood, elephant tree (*xoop inl*, *Bursera hindsiana*), barite, soapstone, and tuff carvings of reptiles is a relatively recent phenomenon among the Comcáac (Figs. 25–28). While these tourist-oriented carvings still feature design elements and natural history knowledge derived from older traditions, they also embody artistic innovations spurred by changes in Comcáac identity and economy which began to accelerate in the 1960s. For example, although ironwood (*Olneya tesota*) was traditionally used by the Comcáac in numerous ways historically, its products became an economic mainstay of Punta Chueca and Desemboque only in the 1970s, when fisheries began to decline so dramatically that Seri men were pressed to find other means of gaining cash income.

Carvings of reptiles soon became some of the most popular tourist crafts sold to American and Mexican visitors to the central Gulf coast. The very first carvings, however, may not have been of animals at all. One version of the origins of the ironwood carving industry holds that around 1962 Desemboque resident José Astorga made bush pilot Alexander "Ike" Russell an amorphous paperweight to keep all his papers from flying off the dashboard of his plane whenever he landed. Another version suggests that a seasonal visitor to Puerto Libertad, a Mrs. Derwin, commissioned Desemboque dwellers to make carv-

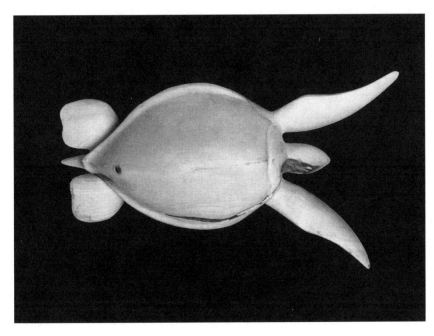

Figure 25. A Green Sea Turtle carved from elephant tree wood by Armando Torres. (Photo Helga Teiwes, 1998)

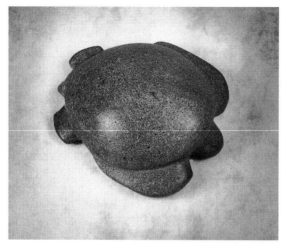

Figures 26–28. Sea turtles carved in stone by Armando Torres: *above*, a Hawksbill; *top left*, a Pacific (Olive) Ridley; *top right*, a Leatherback. (Photos David Burckhalter, 1999)

ings at about the same time. But the favorite story among the Astorga family and other Desemboque residents is that José Astorga had a vision—one gained through celestial travel with extraterrestrial beings—that by initiating a new income-generating activity based on ironwood art forms the Comcáac would be able to survive in the modern world (Sheridan 1997; Mellado n.d.). The Astorga daughters still visit the first ironwood tree from which their father harvested dead branches to elaborate a carving—a tree that still has plenty of harvestable wood within its canopy, which snakes and lizards continue to rely on as habitat.

By the mid-1970s, more than 180 Seri individuals had tried their hand at fashioning animal figurines from dead ironwood, and images of sea turtles, Desert Tortoises, and rattlesnakes had achieved considerable popularity in the international folk arts market (St. Antoine 1994).

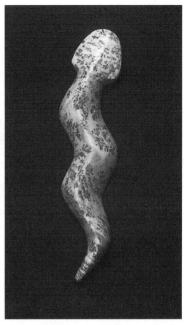

This cottage industry proved so lucrative, providing hundreds of thousands of dollars per year of cash income, that soon non-Seri carvers began to emulate their designs, using machine tools to dramatically increase production. By 1990 perhaps as many as three thousand non-Seri were making ironwood carvings with electric saws, sanders, and buffers, and many of these imitations were misleadingly labeled "hand-made by the Seri Indians" in order to lure unwary customers seeking authentic Indian arts and crafts. Few of these artisans, however, managed to truly capture the feel of a sea turtle swimming or a Sidewinder slithering, as the Comcáac themselves did—perhaps because they lacked any deep familiarity with the animals themselves (Figs. 29 and 30).

In 1991 the Ironwood Alliance was formed, bringing together seventy Seri signators and several conservation groups. The goal was to protect not only the rapidly diminishing ironwood habitat and its wildlife but also the intellectual property rights of the Comcáac (Nabhan and Carr 1994). One conservation strategy the alliance promoted was the use of materials other than slow-growing ironwood—including palo blanco, barite, soapstone, and metamorphosed tuff. In addition, a distinctive "Seri look" would

Figure 29 (*left*). A reptile (possibly a Side-blotched Lizard) "emerging" from a stone base, carved by members of the Robles Barnet family. (Photo Helga Teiwes, 1998)

Figure 30 (*above*). A soapstone Sidewinder. Carver unknown. This and the previous figure show some of the new carving materials and styles that emerged from Comcáac participation in Ironwood Alliance efforts to reduce ironwood cutting and market Seri carvings for the knowledge they display of desert wildlife. (Photo Helga Teiwes, 1998)

Figure 31. A pamphlet distributed by the Ironwood Alliance to aid Seri artisans. (Photo Gary Nabhan, 1992)

be elaborated, which could be merchandised along with associated cultural information; this would give the crafts collective members a distinctive market niche (Fig. 31). It was decided to feature, whenever possible, native animals that the native carvers knew better than anyone else, so that the linkage between culture and natural history could be promoted as one of the values of authentic Comcáac carvings. Dozens of newspaper and magazine articles were published, and thousands of pamphlets and posters were distributed in Mexico and the United States to help promote Seri animal carvings and to discourage machine-made "fake Indian" ironwood crafts.

On behalf of the Arizona-Sonora Desert Museum, I commissioned from expert Comcáac artisans a series of carvings of reptile species. I wanted to demonstrate to them that their knowledge of these animals as expressed through traditional handicrafts could attract sufficient interest to sustain a market niche for which non-Seri could not compete (Figs. 32 and 33).

The images that illustrate this book are testimony to the intimate familiarity many Seri artisans have of the morphology and behavior of the marine and desert fauna of their homelands. Unlike most non-Seri carvings produced in Kino Viejo, Puerto Peñasco, Sonoyta, Hermosillo, or Miguel Alemán/Calle Doce, in which it is impossible to tell which species of turtle or tortoise is being represented, Comcáac carvings include details essential for distinguishing Desert Tortoises from sea turtles, and within sea turtles, Hawksbills from Leatherbacks from Pacific Ridleys. Zebra-tailed Lizards are carved with their tails curved into an upward arch, whereas other *haquímet* are carved with their tails down. Spiny-tailed iguanas can be distinguished from chuckwallas by their manelike crests, and Sidewinders from other rattlesnakes by their horns and lateral slither.

Figure 32 (*left*). Gary Nabhan with Seri carvers. (Photo Thomas Lowe Fleischner, 1992)

Figure 33 (*above*). An early attempt at a Gila Monster in elephant tree wood by Solorio Martínez. (Photo Helga Teiwes, 1998)

"A few days ago, we were all napping under the big salt cedar by the beach. My grandson Esteban woke up before us and went over to play on the beach. He came back running, yelling, 'Grandma I found a sea turtle in all the algae, come and see it.' At first we didn't believe him, but when we finally went over, there it was. When we grabbed it, at first it didn't even move, but it was alive. We picked it up by its carapace. My son-in-law Ricardo carved Esteban's name in the carapace and decided to let it go back into the water even though my brother Gonzalvo suggested we eat it. We overruled him, and took it back to the water. Esteban gave it some of his saliva, which gives good luck to the child. Everyone went along, wading out to let him go."

AMALIA ASTORGA, Desemboque

HOWEVER DILIGENTLY I may try to quantify the relative beauty, utility, and danger of the reptilian world as perceived by the Comcáac, I am ceaselessly aware that these values are ultimately intangibles. What motivates a child to run and tell his parents and grandparents that he is thrilled by the sight of a sea turtle is not something that can be measured. I have no idea how many Comcáac fathers have carved the names of their sons in a turtle's carapace and then released it, but I think of that practice every time I see a sea turtle surface. And even if the Comcáac were never to offer their crafted reptile designs to outsiders, they would still have the means to express the sense of beauty that these creatures bring into their daily lives.

It is much easier for me to understand what Seri individuals tell me about the live-captured snakes I show them than for me to comprehend the serpents that they say inhabit their dreams, songs, and stories as they do sacred springs and ocean passages. These may be what E. O. Wilson (1998) calls "dream serpents," what Carl Jung called "archetypal snakes," or what Joseph Campbell called "mythical beasts": the wild elements still alive inside us (Mundkur 1983; Morris and Morris 1965). While even agnostics might agree that these creatures do play some role in human culture, they still might deny that they play any "real" role in human nature. However, evolutionary biologists and psychologists are now reconsidering this possibility. It may well be that images of serpents in dreams, stories, and songs have an adaptive function, bringing us more quickly to fear, fascination, or readiness-for-action whenever we encounter a snake in our path. The idea that ophidiophobia and biophilia may be genetically "hard wired" remains only a working hypothesis, subject to testing, revision, and perhaps rejection. Further consideration of the rich Comcáac interactions with reptiles may help us better understand to what extent all of humankind is predisposed to strong emotional and physiological responses to seeing or hearing snakes close at hand.

As human evolution scholars Ramona and Desmond Morris (1965) admitted, the riddle of reptiles' primacy in the human psyche has not yet been solved: "No other single animal form has played such an important or varied role in man's thinking. On the one hand, [the reptile, especially the snake] has been a symbol of procreation, health, longevity, immortality, and wisdom; and on the other, it has represented death, disease, sin, lechery, duplicity, and temptation. It is a paradox."

what eats from the turtle's shell, what the turtle eats

Comcáac Perceptions of Local Ecological Interactions

What gets to count as interesting scientific questions depends in part on the interests different cultures and subcultures have in learning about [the aspects of] nature's regularities . . . to which they are exposed. In "the same" environment, different groups . . . will be interested in different questions. Thus, it is interesting to read the history of science as a history of local interests.

SANDRA HARDING, *Is Science Multicultural?* (1998)

I was snorkeling in the middle of the Pacific when I first saw the way sea
turtles attract other lives. I had just passed over a jagged volcanic reef alive
with coral, crabs, and sea anemones. I drifted out over a sandier subtidal
zone, twenty to twenty-five meters deep. One last school of reef fish turned
and flashed their colorful fins, and then the scene opened up and I seemed
to swim alone for a few minutes, immersed in a rich blue-green dream state.

Then, out of the corner of my mask, the dark silhouette of a large fish, or
of my own flipper-laden shadow, appeared between me and the seafloor.
Realizing that the sun was at the wrong angle for the shape to be my own
shadow, I turned my head for a better look.

It was an immature Green Sea Turtle, loping along at the same speed
at which my flippers were moving me. The youngster stretched its front
flippers up, fanned them down, and extended its neck from its carapace.
With an effortless kick from its back flippers it glided forward, then upward. Repeating this sequence several more times, it came up alongside
me, turning its head to look directly into my eyes. Then it breached for
a breath before plunging into the depths once again.

As I followed the turtle into the shadows of a large volcanic boulder, I realized that we were not alone. There, perched on the crags and resting in the crannies of the boulder, were four other turtles, hardly moving except when a surge of water displaced them from where they otherwise seemed anchored in place. This was not merely a family of sea turtles; I saw an entire menagerie, a community of creatures all sharing the resources around this rock. There were multicolored fish touching down on the carapaces of the larger sea turtles, though their colors were hard for me to discern among the shadows. They appeared to be kissing the carapaces, but they were actually scraping something off of them. I cannot presume to know the feeling of a sea turtle being groomed or grazed by another species, but it didn't seem to mind the attention.

When I had surfaced and taken off my gear, I asked others if they had ever seen such an interaction. They had not, but none of us were experts on fish or turtle behavior. I found it hard to imagine that scientists or naturalists had not recorded such interspecific behavior, so easily observable even by a total stranger to the underwater world.

As I later learned from several marine natural history books, such behavior is well documented, but not for all parts of the Pacific. There is a group of fish commonly known as the cleaner wrasse, brilliantly yellow with dark arching stripes and thick, versatile lips. They set up "cleaning stations" to which certain fish, as well as sea turtles, come to have algal muck and ectoparasites removed from their bodies.

The next time I visited Desemboque, I brought along a book picturing various Sea of Cortés reef fishes. If anyone had ever seen this behavior in the midriff of the Gulf, I guessed it would be Guadalupe López, who was respected by many of the Comcáac fishermen for his turtling skills. I found him where we had watched the release of a Green Sea Turtle migrant from a net the spring before; we sat down beneath the nearest mesquite, and I opened the field guide and described what I had seen the day I went snorkeling.

Thoughtfully, he scanned each picture for a moment, passing right by the closest kin to cleaner wrasse that is found in the midriff of the Gulf. He tipped his cowboy hat back a little, exposing his silver-black braids. He shuffled through a few more pages, then paused at one photo, glancing at it from different angles. Pointing to it, he asked me if the words printed below it said that it was a *perico.*

"A *perico,* yes, that's what it says," I answered.

"I'm not sure I've seen what you described to me," he said tentatively, "but this fish, the one we call *perico,* it does something like that. In the early spring, the bottom-touching turtles that have slept and grown a greenish wool of algae on their backs all winter begin to get active, and that's when we see them surface. About that time, the *perico* comes to them, scraping off their blanket of wool. They let them do that, because they don't need to be blanketed from the cold anymore . . ."

Now, whenever I dream of sea turtles swimming, they are not alone. There are fish moving their lips across the carapaces, "shearing" their wool. And there is the quiet voice of Guadalupe narrating the dream, as if he has been diving and swimming with the sea turtles for a very long time.

SOMETIME during the 1980s, certain environmental and human rights activists began to refer to indigenous peoples in general, and Native Americans in particular, as "the first ecologists" (Redford 1989; Buege 1996). Although the term has been used romantically to suggest that all indigenous peoples are "ecologically noble savages" (Buege 1996), there is also another, more positive, implication embedded in this phrase: that since time immemorial, some indigenous people have devoted considerable time and effort to recognizing, understanding, managing, or protecting the ecological interactions among different species, as well as between species and their habitats (Berkes 1999).

Putting aside other aspects of "the first ecologist" stereotype, inquisitiveness about natural interactions is a trait found to varying degrees among all human cultures, indigenous or otherwise. As detailed accounts from indigenous communities in the temperate rain forests of British Columbia (Turner 1996), the deserts of Arizona (Rea 1997), and the lowland tropics of coastal Mexico

(Vásquez-Dávila 1997) have demonstrated, no matter where a society makes its home, some people always take intense interest in interspecific relationships, seeking out and celebrating what ecologist John Thompson (1996, 1999) calls "the interaction diversity" of the myriad lives surrounding them.

"The tortoise is an easily pleased animal, [even though] he cannot comprehend either the Mexican or the American. Because of this [innocence], he is content to walk under branches until he comes upon a heap of edible greens, which he eats before he continues on."

LYDIA IBARRA FÉLIX, *Punta Chueca, explaining a Comcáac song*

While I do not wish to imply that all indigenous cultures, or even all Comcáac community members, have shared certain intellectual interests with "ecologists" across time, I have heard the many ways contemporary Seri individuals discuss the interaction of reptiles with other animals, plants, or their habitats. I hope to describe this domain of traditional ecological knowledge, without, however, reinforcing the purely romantic take on the concept of "first ecologists" or implying that this stereotype readily fits Comcáac culture. Instead, I wish to demonstrate how the Comcáac express their "indigenous scientific knowledge" of ecological interactions, using means that are often qualitatively very different from those routinely used by professional, academically trained scientists (Nabhan 2001).

In so doing, I follow the urging of folklorist Rayna Green (1981) of the Smithsonian: we should define the natural sciences broadly so that indigenous science is granted validity, rather than simply assuming that science is "the exclusive property of any one group, [for] every group has a 'science' based on observation, experiment and tradition . . . whether scientific knowledge is developed through 'experiment' or 'trial and error observation,' and whether knowledge is organized into a 'systematic body' or 'diffused throughout a range of cultural expression and behavior.'" For indeed, certain Comcáac seafarers and academically trained field ecologists show great interest in some of the very

Figure 34. A rattlesnake consuming a Spadefoot Toad, as interpreted by José Luis Blanco, who also carves rattlesnakes eating horned lizards and horned lizards "escaping" by wounding rattlesnakes. (Photo Helga Teiwes, 1998)

same ecological interactions, and each group offers valid insights into how those interactions work (Fig. 34). Both may recognize interactions between predators and prey, herbivores and herbs, plants and pollinators, or ectoparasites and their hosts (Nabhan 2000a, 2000b). They may be equally curious about competitors, mutualists, and cohabitants of the same cavity, cave, nest, or water hole. Both may interpret contrasting situations found in nature as "natural experiments," a means of developing hypotheses and theories to explain why one kind of animal seems to thrive in some conditions but not in others.

In 1976, ethnobotanists Robert Bye and Maurice Zigmond coined the term *ethnoecology*, an eminently useful term that they, however, restricted to "the area of [scientific] study that attempts to illuminate in an *ecologically* revealing fashion man's interactions with and relationship to his environment" (Bye and Zigmond 1976). I wish to use the term in a broader sense: as any community's attempts to understand other species' interactions in their own cultural terms (Fig. 35). In this sense, ethnoecology is what any cultural community knows of the diversity of ecological interactions involving humans and other life-forms—animals, plants, microbes, and their habitats (Nabhan 2001). In addi-

Figure 35. Some Seri individuals who retain knowledge of herbivore-plant interactions use these insights to gather plant foods once their normal season has passed. Here, Amalia Astorga and Angelita Torres excavate a packrat midden to obtain mesquite pods gathered earlier in the season by the packrat (*Neotoma* sp.). In a sense, they "short-circuit" food chains, harvesting the animal's energy supply even when they fail to capture the animal itself. (Photo Christine Keith, 1997)

tion, it incorporates the community's own influences, conscious and unconscious, on these relationships.

"*Hehe iti cooscl?* That's the lizard that lives up in trees. In mesquites. Back in the mountains [nodding eastward]. On the desert flats as well. It eats ants during the hottest time of the year. It too is eaten, by Red-tailed Hawks. My mother knew its song."

JOSEFINA IBARRA FÉLIX, *Punta Chueca, speaking on tree lizards*

One of the ways I gain an initial impression of a culture's knowledge of ecological interactions is by looking at their lexicon. Any dictionary, whether written or oral, is full of connotations and inferences regarding interspecific relationships and habitat preferences—where an animal lives, what it eats, or with what other species it interacts. Many of these are implicit in the very names of plants and animals. When such a name marks an organism's relationship to another organism or to a particular habitat, I describe it as an "ecological referent."

The language of the Comcáac, Cmique Iitom, appears to have more ecological referents in its biosystematic lexicon than do those of neighboring desert peoples (Nabhan 2000b). Not only that, but with regard to reptiles specifically, ecological referents shape the Comcáac lexicon almost overwhelmingly, virtually shutting out other referents such as sound symbolism. By sound symbolism, I refer to the linguistic sounds—of birdsong, for example—or the tactile dimensions of movements—of a fish or snake, say—in shaping the zoological names (Berlin 1992). This process in turn bestows special aural and psychological significance on the resulting names (Hinton, Nichols, and Ohala 1994).

I have attempted to search for sound symbolism embedded in Comcáac names for reptiles, wondering, for example, if the word for rattlesnake, *cocázni*, onomatopoetically echoes the sound of a rattle. To no avail, however. Any aural patterns in the biosystematic lexicon appear to be minor compared to the many references to the habitats where reptiles dwell, the plants that occur with them, and the morphological similarities between land and marine reptiles.

How do such linguistic considerations reflect Comcáac ecological knowledge? Table 18 lists some of the names for reptiles that incorporate "habitat markers"—such as geomorphological settings or rock, water, and soil types—in their compound lexemes. These names, bearing meanings like Gila Monster of the high seas, chuckwalla-like mountain dweller, and small lizard in the tree, can be considered akin to Marsh Wren, Canyon Towhee, and Alkali Sunflower in common English.

TABLE 18

Habitat Markers Imbedded in Comcáac Names for Reptiles

Species	Comcáac Name	Habitat Alluded To
TURTLES AND TORTOISES		
KINOSTERNIDAE		
Kinosternon spp.	xtamáaija	mud, lagoon, estuary?
LIZARDS		
CROTAPHYTIDAE		
Crotaphytus spp.	hast coof	rocky uplands
Gambelia wislizenii	hantpízl	silty soil?
PHRYNOSOMATIDAE		
Phrynosoma solare	hant coáaxoj	silty soil?
SNAKES		
COLUBRIDAE		
Chionactis occipitalis	hant quip	sand dunes
HYDROPHIIDAE		
Pelamis platurus	xepe ano cocázni	seas
CROCODILES		
CROCODYLIDAE		
Crocodylus acutus	xepe ano heepni, xepe ano paaza	seas

"A Desert Tortoise was once launched way up in the sky. It had been traveling very slowly on the ground, but in the sky it wanted to go faster."
FRANCISCO "CHAPO" BARNET OF PUNTA CHUECA, *explaining the absurd humor of a song about a Desert Tortoise let loose from rocky ground*

In addition, a number of Comcáac names for plants encompass the names of reptiles as part of compound lexemes (Table 19). For instance, the yellow-flowered shrub *Trixis californica* is known by the Comcáac as *cocaznóotizj*, "rattlesnake's foreskin." This name, a bawdy morphological metaphor, plays on the shape of the plant's composite flower with its subtending bracts, likening it to the "foreskin" of the penis-shaped serpent. The Comcáac also draw upon metaphors of texture and color, likening the mottling on the *bebelama* fruit of *Sideroxylon occidentalis*, for example, to the skin patterns of Gila Monsters.

> "*Paaza? Paaza?* What do you mean, the animal or the fruit? The tree is named that way because of the way its fruit looks like the skin of the Gila Monster. Both have little black *puntitas* [tapping her hand with her finger, as if marking it with a pen point]."
>
> RAMONA BARNET, *with Mercedes, her sister-in-law, in Punta Chueca*

The Comcáac also find a resemblance between the shape of Boa Constrictors and that of the slender snaking stems of the Sina Cactus (*Stenocereus alamosensis*), as reflected in the shared, polysemous name *xasáacoj* (Felger and Moser 1985). Although Boa Constrictors are seldom seen today, this analogy reminds the Comcáac that their people have historically observed Boa Constrictors within their aboriginal territory. Curiously, it was not until the 1950s that biologist Chuck Lowe (1959) recognized that Boa Constrictors occur north of Guaymas and west of Hermosillo, in many of the same places that the Sina Cactus reaches its northwestern limits—well within the historic Comcáac range.

There are, of course, reasons other than ecological significance for referring to a plant using an animal name, or vice versa. For example, the construction *caay ixám*, a name for the ornamental hard-shelled gourd *Cucurbita pepo*, means "horse's gourd," even though horses do not necessarily eat it (Felger and Moser 1985); both, however, are domesticated and are jokingly considered monstrosities by some Seri individuals. Another Comcáac name, *moosni iha*, "sea turtle's belongings," is used for two land plants, *Palafoxia arida* and *Dithyrea californica*. The reference to sea turtles does not imply an ecological interaction; instead it reflects the tradition of placing freshly butchered sea turtle meat on a mound of these herbs to keep the meat from getting covered with sand.

TABLE 19

Plants and Algae with Reptile-Related Names in Cmique Iitom

Species	Comcáac Name	Loose English Translation	Plant/Reptile Relationship	Source
PLANTS				
ASTERACEAE				
Chaenactis carphoclinia	xtamoosnóohit	"Desert Tortoise's forage"	forage	Felger and Moser 1985; Nabhan field notes
Palafoxia arida	moosni iha, moosni oohit	"sea turtles' belongings," "sea turtle's forage"	no known relation currently understood	Felger and Moser 1985; Reina-Guerrero and Morales 1998; Nabhan field notes
Trixis californica	cocaznóotizj	"rattlesnake's foreskin"	morphological metaphor	Felger and Moser 1985; Nabhan field notes and specimens deposited in University of Arizona Herbarium, cited in Nabhan 2000b
Verbesina palmeri	cocaznóotizj caacöl	"rattlesnake's foreskin"	morphological metaphor	Felger and Moser 1985
BRASSICACEAE				
Dithyrea californica	moosni iha	"sea turtle's belongings"	no known relation	Felger and Moser 1985
CACTACEAE				
Stenocereus alamosensis	xasáacoj	"Boa Constrictor"	morphological metaphor	Felger and Moser 1985
CAMPANULACEAE				
Nemacladus glanduliferus	xtamáaija oohit	"mud turtle's forage"	forage	Felger and Moser 1985; Nabhan field notes
EUPHORBIACEAE				
Croton californicus	moosni iti hatéepx	"sea turtle (meat's) cloth" or "resting place"	"place mat" for newly butchered turtle meat	Felger and Moser 1985; Nabhan field notes
HYDROPHYLLACEAE				
Phacelia ambigua	xtamoosnóohit	"Desert Tortoise's forage"	forage	Felger and Moser 1985; Nabhan field notes
NYCTAGINACEAE				
Abronia maritima, *A. villosa*	moosni iti hatéepx	"sea turtle (meat's) cloth" or "resting place"	"place mat" for newly butchered turtle meat	Nabhan field notes

Taxon	Seri name	Gloss	Category	Reference
SAPINDACEAE				
Sideroxylon occidentale	paaza	"Gila Monster"	morphological metaphor	Nabhan field notes
ZYGOPHYLLACEAE				
Fagonia californica, *F. pachyacantha*	xtamoosnóohit	"Desert Tortoise's forage"	forage	Felger and Moser 1985; Nabhan field notes
MARINE ALGAE				
BONNEMAISONIACEAE				
Asparagopsis taxiformis	moosníil ihaquéepe, moosni oohit	"blue sea turtle's delight," "sea turtle's forage"	forage	Felger and Moser 1985
GRACILARIACEAE				
Gracilaria textorii	moosni yazj	"sea turtle's covering"	inquiline (or bioassociate)	Felger and Moser 1985
HALYMENIACEAE				
Cryptonemia obovata	moosni ipnáil	"sea turtle's skirt"	habitat indicator, inquiline (or bioassociate)	Felger and Moser 1985
Halymenia coccinea	moosni ipnáil	"sea turtle's skirt"	habitat indicator, inquiline (or bioassociate)	Felger and Moser 1985
RHODYMENIACEAE				
Padina durvillaei	moosni ipnáil, moosni yazj	"sea turtle's skirt," "sea turtle's covering"	habitat indicator, inquiline (or bioassociate)	Felger and Moser 1985

TABLE 20

Comcáac Recognition of Functional Relationships
between Certain Reptiles and Other Species

Reptile Species	Plant/Algae Species	Relationship
TURTLES AND TORTOISES		
TESTUDINIDAE		
Gopherus agassizii	*Allionia incarnata*	forage plants
	Chaenactis carphoclinia	for Desert Tortoise
	Chorizanthe brevicorun	
	Fagonia californica	
	Fagonia pachyacantha	
	Phacelia ambigua	
	Trianthema portulacastrum	
KINOSTERNIDAE		
Kinosternon sonoriense	*Nemacladus glanduliferus*	forage plant
		for mud turtle
CHELONIIDAE		
Chelonia mydas	*Asparagopsis taxiformis*	algal forage plants
	Cryptonemia obovata	for Green Sea Turtle
	Dithyrea californica	
	Gracilaria textorii	
	Rhodymenia hancockii	
	Cryptonemia obovata	carapace-covering algae
	Halymenia coccinea	for Green Sea Turtle
	Padina durvillaei	
	Rhodymenia hancockii	
SNAKES		
COLUBRIDAE		
Oxybelis aeneus	*Jatropha cinerea*	microhabitat, camouflage
		for Brown Vinesnake

Table 20 details the reptile names that truly allude to functional interspecific relationships among creatures. Some of these Comcáac names seem to me to be particularly insightful about ecological relationships—ones, indeed, I have seldom heard discussed by academically trained naturalists and field biologists.

For example, *hamísj catójoj*, the name for what English speakers know as the Brown Vinesnake (*Oxybelis aeneus*), refers to this snake's mimicry of the stems of Ashy Limberbush (*Jatropha cinerea*), on which it hides. This name makes perfect sense (and is therefore quite memorable) in a hyperarid land where few true vines even grow.

Comcáac knowledge of ecological interactions extends beyond the relationships encoded in the names of animals and plants. Tables 21 and 22 contrast Western scientific and Comcáac traditional observations on the habitat preferences of various reptiles.

For instance, Eva López once sang a song about a whiptailed lizard mother warning her children not to run on a path frequented by rattlesnakes. Later the mother sees a rattler with the tail of one of her babies in its mouth. This song is as much about maternal care in humans as it is about rattlesnake-whiptail predation, and its touch of black humor is a potent reminder to small children that they should heed the warnings of their mothers.

"The eelgrass—it's a special food of the sea turtles both here in the Canal del Infiernillo and in the high seas. But it is being disturbed by the crab traps: as the wire in metal cages oxidizes, it burns the hatchlings of many species."
ERNESTO MOLINA, *Punta Chueca*

Or then there are the fish we have already encountered that clean the "woolen blankets" of marine algae off the carapaces of winter-dormant Green Sea Turtles. According to Felger and Moser (1989), this algal mat includes the red alga *Gracilaria textorii*, which the Comcáac call *moosni yazj*, "sea turtle's membranes." It grows up to 30 centimeters on the carapaces of the partially buried, overwintering turtles (Felger, Cliffton, and Regal 1978). Although Guadalupe López carefully attributed the algae-cleaning behavior to *pericos*, Bumphead Parrotfish (*Scarus perrico*), the standard work on rocky reef fishes of the Sea of Cortés lacks any reference to *pericos* setting up cleaning stations for sea turtles (Thomson, Findley, and Kerstitch 1987). Nevertheless, one of its authors, ichthyologist Donald

TABLE 21

Reptile Habitats on the Sonoran Coast and Adjacent Islands, as Recorded by Western Scientists

Species	Rocky Uplands	Silt/Sand/Gravel Alluvial Valleys	Dunes and Sandy Beaches	Freshwater and Estuarine	Marine
TURTLES AND TORTOISES					
TESTUDINIDAE					
Gopherus agassizii	•				
KINOSTERNIDAE					
Kinosternon flavescens				•	
K. sonoriense				•	
CHELONIIDAE					
Caretta caretta					•
Chelonia mydas					•
Eretmochelys imbricata					•
Lepidochelys olivacea					•
DERMOCHELYIDAE					
Dermochelys coriacea					•
LIZARDS					
CROTAPHYTIDAE					
Crotaphytus dickersonae	•				
C. nebrius	•				
Gambelia wislizenii	•	•			
IGUANIDAE					
Ctenosaura conspicuosa	•				
C. nolascensis	•				
Dipsosaurus dorsalis	•	•	•		
Sauromalus hispidus	•				
S. obesus	•				
S. varius and S. varius × S. obesus × S. hispidus	•	•			
PHRYNOSOMATIDAE					
Callisaurus draconoides		•	•		
Holbrookia maculata		•	•		
Phrynosoma solare	•	•			
Sceloporus clarkii	•				
S. magister	•	•			
Urosaurus graciosus		•			
U. ornatus		•		•	
Uta stansburiana	•	•	•		

TABLE 21 (continued)

Species	Rocky Uplands	Silt/Sand/Gravel Alluvial Valleys	Dunes and Sandy Beaches	Freshwater and Estuarine	Marine
EUBLEPHARIDAE					
Coleonyx variegatus	•	•	•		
GEKKONIDAE					
Phyllodactylus xanti	•				
TEIIDAE					
Cnemidophorus burti		•			
C. tigris	•	•			
XANTUSIIDAE					
Xantusia vigilis			•		
HELODERMATIDAE					
Heloderma suspectum	•	•			
SNAKES					
BOIDAE					
Boa constrictor		•		•	
Charina trivirgata	•			•	
COLUBRIDAE					
Arizona elegans		•			
Chilomeniscus stramineus		•	•		
Chionactis occipitalis	•	•	•		
Hypsiglena torquata	•	•			
Lampropeltis getula	•	•			
Masticophis bilineatus	•	•		•	
M. flagellum	•	•			
M. slevini	•	•			
Oxybelis aeneus	•	•			
Pituophis melanoleucus	•	•			
Rhinocheilus lecontei	•	•			
Salvadora hexalepis		•	•		
Trimorphodon biscutatus	•				
ELAPIDAE					
Micruroides euryxanthus	•	•			
HYDROPHIIDAE					
Pelamis platurus				•	•
VIPERIDAE					
Crotalus atrox	•	•			

Continued on next page

TABLE 21 (continued)

Species	Rocky Uplands	Silt/Sand/Gravel Alluvial Valleys	Dunes and Sandy Beaches	Freshwater and Estuarine	Marine
C. cerastes	•	•	•		
C. estebanensis	•	•			
C. molossus	•	•			
C. scutulatus	•	•			
C. tigris	•	•			
CROCODILES					
CROCODYLIDAE					
Crocodylus acutus				•	•

Sources: González-Romero and Alvarez Cardenas 1989; Grismer 1999.

Thomson, assumes that Guadalupe's commentary probably describes with accuracy a behavior that scientists have yet to observe (pers. comm.).

> "You search for the *moosni hant coit* [winter-dormant sea turtles] by the greens
> on their carapaces, since you could see only a small portion of the entire shell itself.
> They could be in straits that are fairly deep, but you need clear water to see them,
> and a long harpoon, seven arm-lengths long. . . . [Then] when you first put sea
> turtle meat in your mouth, you have the taste of the greens they have eaten."
> RAMÓN PERALES, *Punta Chueca*

There are some discrepancies between the forage preferences of herbaceous reptiles as reported by Comcáac observers and the forage inventories compiled by academically trained field ecologists (Van Devender et al. 2002)—differences due, quite likely, to location-by-location variation in diets rather than "errors" on the part of either set of observers. For instance, the Comcáac have observed Desert Tortoises eating seven native plants both on Isla Tiburón and on the central Gulf coast of the Sonoran mainland. Of these, two are documented in the analyses of Desert Tortoise scat and foraging behavior in other desert localities.

TABLE 22

Reptile Habitats on the Sonoran Coast and Adjacent Islands, as Observed by Seri Consultants

Species	Rocky Uplands	Silt/Sand/Gravel Alluvial Valleys	Dunes and Sandy Beaches	Freshwater and Estuarine	Marine
TURTLES AND TORTOISES					
TESTUDINIDAE					
Gopherus agassizii	•	•			
KINOSTERNIDAE					
Kinosternon flavescens				•	
K. sonoriense				•	
CHELONIIDAE					
Caretta caretta					•
Chelonia mydas					•
Eretmochelys imbricata					•
Lepidochelys olivacea			•	•	
DERMOCHELYIDAE					
Dermochelys coriacea					•
LIZARDS					
CROTAPHYTIDAE					
Crotaphytus dickersonae	•				
C. nebrius	•				
Gambelia wislizenii		•	•		
IGUANIDAE					
Ctenosaura conspicuosa	•				
C. nolascensis	•				
Dipsosaurus dorsalis	•	•			
Sauromalus hispidus	•				
S. obesus	•				
S. varius and S. varius × S. obesus × S. hispidus	•				
PHRYNOSOMATIDAE					
Callisaurus draconoides		•	•		
Phrynosoma solare		•	•		
Sceloporus clarkii	?	?			
S. magister	•	•			
Urosaurus graciosus			•		
U. ornatus			•		

Continued on next page

TABLE 22 (continued)

Species	Rocky Uplands	Silt/Sand/Gravel Alluvial Valleys	Dunes and Sandy Beaches	Freshwater and Estuarine	Marine
Uta stansburiana			•		
EUBLEPHARIDAE					
Coleonyx variegatus	•	•	•		
GEKKONIDAE					
Phyllodactylus xanti	•				
TEIIDAE					
Cnemidophorus burti		•			
C. tigris		•			
XANTUSIIDAE					
Xantusia vigilis			•		
HELODERMATIDAE					
Heloderma suspectum	•	•			
SNAKES					
BOIDAE					
Boa constrictor				•	
Charina trivirgata	•	•			
COLUBRIDAE					
Arizona elegans		•	•		
Chilomeniscus stramineus			•		
Chionactis occipitalis		•	•		
Hypsiglena torquata	•				
Lampropeltis getula		•	•		
Masticophis bilineatus	•	•			
M. flagellum	•	•			
M. slevini	•				
Oxybelis aeneus		•			
Pituophis melanoleucus	•	•	•		
Rhinocheilus lecontei		•			
Salvadora hexalepis		•	•		
Trimorphodon biscutatus	•				
ELAPIDAE					
Micruroides euryxanthus		•	•		
HYDROPHIIDAE					
Pelamis platurus					•
VIPERIDAE					
Crotalus atrox	•	•	•		

TABLE 22 (continued)

Species	Rocky Uplands	Silt/Sand/Gravel Alluvial Valleys	Dunes and Sandy Beaches	Freshwater and Estuarine	Marine
C. cerastes		•	•		
C. estebanensis	•	•			
C. molossus	•	•			
C. scutulatus	•	•			
C. tigris	•	•	•		

CROCODILES

CROCODYLIDAE

| Crocodylus acutus | | | | • | • |

Sources: González-Romero and Alvarez Cardenas 1989; Grismer 1999.

Another one, *Chaenactis carphoclinia*, is in the same genus as plants that herpetologist Tom Van Devender and colleagues (in press) have documented from tortoise scat in southern Arizona, but C. *carphoclinia* itself does not occur on their study sites. A fourth species cited by the Comcáac as tortoise forage, *Chorizanthe brevicornu*, was recently noted as a significant food plant for tortoises at some Arizona localities (Van Devender et al. 2002). Another species described by the Comcáac, *Trianthema portulacastrum*, a false purslane, is so similar in growth form to the purslane herbs known from many tortoise diets that Van Devender and I would consider it a probable component in Sonoran tortoise diets. Two additional tortoise forages, an *Allionia* and a *Phacelia*, have not been documented in any Desert Tortoise forage inventories of which I am aware, but this may be related to the geographic specificity of such studies (Nabhan 2002).

"*Xtamoosnóohit* [tortoise's forage]? This plant, it's an herb that comes up in the spring. It looks a little like an onion. The tortoises eat other herbs as well."
JESÚS ROJO, *Punta Chueca*

These examples indicate that local knowledge, though in some ways different, is potentially complementary to and verifiable by the knowledge accrued by field ecologists. In the view of science historian Sandra Harding (1998), it is unlikely that desert ecologists and an indigenous group like the Comcáac would ever come up with identical inventories of tortoise forages, even if both sets of observers were given the same amount of time or field effort to develop their lists. Indeed, according to Harding, academically trained scientists and locally trained indigenous people probably are *unable* to see the world in the same way. "The claim here is not that knowledge based on some local interactions with nature is always more accurate; obviously very often it is not. . . . Rather, the claim is that cultures' different locations in heterogeneous nature expose them to different regularities of nature" (Harding 1998).

What does "regularities of nature" mean? Harding, as a social scientist, may well be stressing the economic needs of indigenous societies in her use of the term. In our case, then, "regularities of nature" might refer especially to the salient patterns of ecological interactions that Seri individuals are most likely to observe as they go about the critical business of hunting, fishing, or gathering. In contrast, academically trained biologists, being unconstrained by the need to realize immediate nutritional or financial gain, might be expected to give their attention to and be impressed by a wider diversity of organisms and interactions, even if they were to visit the same locations that the Comcáac frequent.

And yet, when I consider whether the Comcáac focus only on these ecological interactions involving the most economically valuable species in their midst, I see no obvious bias. Utility alone is not a good predictor of Comcáac interest in the natural world. Some ecological processes may simply be more fascinating to the human mind than others, regardless of their practical value. I have been struck at the number of times a Seri ecotourist guide and foreign naturalists have stood together watching ecological interactions in utter delight: those between whipsnake and rattler, horned lizard and ant, or tortoise and cactus. Academic scientists may ask different questions and answer them using different tools than do local naturalists, but the same sense of biophilia holds when they witness a female sea turtle lumber onto a beach to lay her eggs or a spiny-tailed iguana climb high into the arms of a Cardón cactus.

Plate 1. Isla Alcatraz in Bahía Kino, home to hybrid chuckwallas resulting from cultural dispersal.
(Photo Helga Teiwes, March 1998)

Plate 2. Habitat for hybrid chuckwallas on Isla Alcatraz. (Photo Helga Teiwes, March 1998)

Plate 3. Spiny-tailed iguana atop Cardón cactus, Arroyo Limantur, Isla San Esteban. (Photo Lew Walker, May 1962, courtesy of Arizona-Sonora Desert Museum)

Plate 4 (*right*). Spiny-tailed iguana, Isla San Esteban. (Photo Lew Walker, May 1962, courtesy of Arizona-Sonora Desert Museum)

Plate 5 (*below*). Green Sea Turtle, Sea of Cortés. (Photo Lew Walker, June 1966, courtesy of Arizona-Sonora Desert Museum)

Plate 6 (*above*). Loggerhead Turtle, Sea of Cortés. (Photo Lew Walker, January 1969, courtesy of Arizona-Sonora Desert Museum)

Plate 7 (*left*). Track of a Green Sea Turtle. (Photo Lew Walker, n.d., courtesy of Arizona-Sonora Desert Museum)

Plate 8 (*right*). Jesús Rojo of Punta Chueca singing sea turtle harpooning songs, with Gary Nabhan taking notes. (Photo James Griffith, January 1998)

Plate 9 (*below*). Comcáac storytellers Adolfo Burgos and Amalia Astorga with a Desert Tortoise, Arizona-Sonora Desert Museum. (Photo Helga Teiwes, October 1, 1997)

Plate 10 (*opposite*). Amalia Astorga of Desemboque holding a kingsnake. (Photo Helga Teiwes, October 3, 1997)

Plate 11 (*above*). A speared Green
Sea Turtle, Isla San Esteban.
(Photo #10 by William N. Smith,
September 20, 1947, courtesy of
University of Arizona Special
Collections)

Plate 12 (*right*). The roasted
lower plate of a sea turtle, ready
for eating, Desemboque. (Photo
#17 by William N. Smith, June
24, 1949, courtesy of University
of Arizona Special Collections)

Plate 13 (*left*). Turtle hunting, Punta Perla, Isla Tiburón. (Photo #9 by William N. Smith, October 3, 1949, courtesy of University of Arizona Special Collections)

Plate 14 (*right*). A woman breaking open a turtle at one of the winter hunting camps on Isla Tiburón. (Photo #13 by William N. Smith, n.d., courtesy of University of Arizona Special Collections)

Plate 15 (*above*). A close-up of the head of a pre-ceramic green-stone turtle or tortoise, from the "Ach-aach" shell midden, on the north coast of Isla Tiburón. (Photo #20 by William N. Smith, n.d., courtesy of University of Arizona Special Collections)

Plate 16 (*opposite top*). A prehistoric sandstone carving of a Leatherback Turtle disinterred by Raúl Molina near Punta Chueca, now in the Seri museum in Kino, Sonora. (Photo Gary Nabhan, n.d.)

Plate 17 (*opposite bottom*). A sand drawing on the beach, made by Seri children. (Photo #15 by William N. Smith, 1946–1966, courtesy of University of Arizona Special Collections)

Plate 18. A Comcáac medicinal plant bundle, Desemboque. (Photo Laurie Monti, March 1998)

Plate 19 (*above*). Painting a Leatherback, Isla Tiburón or Punta Chueca. (Photo #18 by William N. Smith, Feb./Mar. 1965, courtesy of University of Arizona Special Collections)

Plate 20 (*left*). A turtle harpoon carved of ironwood by Guadalupe López. (Photo David Burckhalter, 1999)

Plate 21. Four Piebald Chuckwallas. (Photo #13 by William N. Smith, 1949, courtesy of University of Arizona Special Collections)

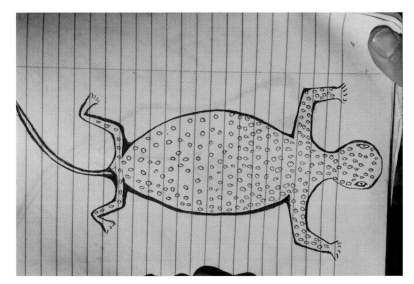

Plate 22 (*left*). Drawing of a Gila Monster by Amalia Astorga. (Photo Gary Nabhan, 1998)

Plate 23 (*below*). A Zebra-tailed Lizard carved in ironwood, showing the characteristic position of its tail. (Photo Helga Teiwes, 1997)

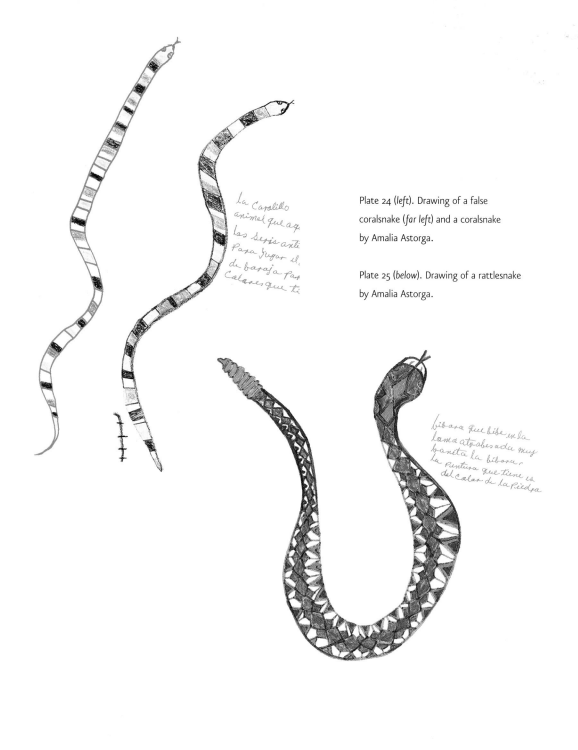

La Caralillo
animal que aq...
las seris anti...
Pasa jugar el...
de baraja par...
Calaresque ti...

Plate 24 (*left*). Drawing of a false
coralsnake (*far left*) and a coralsnake
by Amalia Astorga.

Plate 25 (*below*). Drawing of a rattlesnake
by Amalia Astorga.

Vibara que bibe en la
loma atrabesada muy
banita la bibora
La Pintura que tiene es
del calor de la Piedra

"A coyote tried to eat a Desert Tortoise, but it didn't have any luck. The tortoise tucked its entire body into its shell, and all the coyote could do was roll it around like a stone."

JESÚS ROJO, *Punta Chueca*

THE INTERSPECIFIC RELATIONSHIPS the Comcáac discuss the most are undoubtedly those between "pets" and people. By "pet," here, I mean any animal that is kept within a household for any period of time, or is tended, fed, and played with, regardless of whether it is wild or domesticated. The pet that occupies the most peculiar place in Comcáac culture is the dog, once trained to hunt Desert Tortoises and other wildlife but also symbolically associated with a family's ihízitim and ancestral lineage. Dogs are named for deceased Seri individuals and, according to José Juan Moreno, can be given names only by elderly, post-reproductive men or women (Smith 1951).

Yet dogs are by no means the only animals kept by Comcáac children as pets; horned lizards, chuckwallas, sea turtles, javelina, and Mule Deer have also fallen into that august category. In some cases, these pets have been rescued from the wild when they were young, particularly if they were found wounded or orphaned. One boy raised a Regal Horned Lizard to maturity, taking it out every day before school and placing it near an anthill so that it could live on the same food it would seek if it had been left in the wild. Regardless of whether an animal is kept for a week or for many years, its presence typically generates animated discussion among children and adults alike. Contrary to certain theories about hunter-gatherers' relationships with wildlife (Shepard 1993), these tended pets are valued for their individual personalities, not merely for the archetypal traits associated with their species.

The most extraordinary relationship I have heard of between a Seri adult and an individual animal was between Amalia Astorga and a small lizard that she called Efraín the Side-blotched Lizard (Astorga, Nabhan, and Miller 2001). She describes the relationship in a story that won the *Stepping Stones* magazine award for best multicultural children's book of 2001:

His name was Efraín. He was a Side-blotched Lizard. When I first saw him, he had come close to my house and was looking at me very carefully. . . . He had no tail past this point [midthigh] . . . but he was built like a man, stout, with strong arms and thick legs. He was built like a wrestler, that Efraín.

All in all, he spent seven years with us, visiting with me every day. . . . He would never fail to make a show. . . . For the first five months, he came alone. Then . . . he started to bring his family with him to drink. . . . They learned to drink water [from a saucer] just as he did. He brought with him four lizard children and his wife.

His children all had bright blue-green tails, but his wife was much the same color he was. . . . They all came visiting for many years, but Efraín remained my favorite. My sister said that he was as true to me as if he were my husband. . . . The entire village knew him.

Then one day, when I was out gathering limberbush stems . . . he came to my house when I was gone. . . . The neighborhood dogs [saw him and chased him]. He must have tried to run from them; they caught him by the back leg. . . . Before I returned, he had died.

When I came home, my daughter Ana was crying. *"What shall we do?"* we cried. "We must bury him in the desert as we do a great man." We took him away . . . in the desert and dug into the earth and buried him there. Ana marked and blessed his grave with a cross. For many days I cried and cried. I didn't want to eat. I began to lose weight. I missed Efraín so much.

Since his death, Amalia and Ana have occasionally been visited by lizards that they believe are Efraín's survivors, but never again as frequently as when he was alive. Curiously, about the time I first heard the story of Efraín, the renowned archeologist Julian Hayden invited me to his adobe home in Tucson, where we talked about his trip to Seriland with ethnologist Gwyneth Harrington and O'odham folklorist Juan Xavier more than fifty years ago. We sat out on his patio, and he began to tell me of his visit to the Barnet family camp on Isla Tiburón.

Suddenly, several lizards, including a fat desert spiny, came running up in front of him. "Just a minute," Julian said, "I forgot that it's happy hour." As he went into his house, the lizards retreated. After a couple of minutes, Julian brought out a couple of cold drinks, some crackers, and some cheese. The lizards reappeared, and he began hand-feeding them pieces of the cheese. Julian claimed that they had been doing this with him for years, perhaps over multiple generations, and that the lizards would not accept cheese from anyone else.

Extended relationships between individual wild reptiles and individual humans have seldom been reported, let alone celebrated. They allow us to imagine possibilities for interspecific interactions that most of us seldom consider, and fewer still participate in directly. And they remind us that the web of interactions going on around us is far more complex than Western science has yet to fully explore.

the comcáac as conservationists

Practicing What They Preach, and Benefiting from Alliances

The fact that sea turtles still exist in Mexico today is testimony to their evolutionary heritage, but they won't last long without increased protection. New ideas and alliances must be found if the latest chapter in the sea turtle tragedy is to have a happier ending.

KIM CLIFFTON, "Leatherback Turtle Slaughter in Mexico" (1990)

After a full month's stay on the Sea of Cortés coast, working daily with Seri friends, I made the rounds in both villages to say good-bye to the families who had helped with my work. As we exchanged jokes and stories, I mentioned with excitement that just that morning I had seen young migratory sea turtles returning in groups to the Canal del Infiernillo, swimming together on the edge of the eelgrass beds off Isla Tiburón's eastern shore. While I was telling this to Carolina Hoeffer in Desemboque, her grandchildren came running up, just as excited as I was.

"*Moosni! Moosnáapa!*" they yelled. "Sea turtle! The true turtle!"

From where I stood in Carolina's front yard I saw two boys in the bed of a truck loaded with crates of fish. They were holding a live Green Sea Turtle by its front flippers; it stretched from their waist to their ankles. The turtle's back flippers churned as if it were reaching for water.

We scurried over to the back of the flatbed to get a closer look. All the little boys on the street crawled onto the flatbed and petted the baby sea turtle, squealing with delight. The turtle, unaccustomed to being this far from water, snuggled into the heaps of wet fish. Its lead-colored head no longer glistened like the ones I had seen in the channel earlier that morning.

"Where did you get it?" I asked.

"The fishermen accidentally caught it in their nets," an older man replied. "They rescued it when pulling up their catch. It's here for the kids to see, then we'll take it back to the water and let it loose."

The truckload of men and boys soon drove back to the beach. They kept the turtle tethered until the tide shifted and the water deepened in front of them. The children gathered around it, laughing and talking excitedly in their native tongue until the adults unleashed it and it scurried into the waves.

Alfredo López missed getting a look at this turtle, and later asked me to describe it in detail.

"*Moosnáapa* or *cooyam?*" he asked. "What color was it? And what size?"

I described the features I could remember, and he brooded over them. I watched him while his dark brown eyes seemed to revisualize every sea turtle he had encountered over his sixty-some years of living along the coast. "That's probably a young *moosnáapa*. It's true, the migrants are already coming north in small groups, but this one was found by itself, and its markings don't sound like those of the ones that migrate long distances. They say it was all alone in the nets?"

"That's what they said. Would you typically let it go under those circumstances?"

"That's what the whites don't understand about the way we treat sea turtles," he said quietly, deliberately, as if my question implied that the Comcáac regard for sea turtles remained obscure not only to me but to his Mexican neighbors as well. "The Cocsar or Yoris"—his terms for non-Comcáac—"they see that we still hunt them on occasion, so they assume we do it as they do. Cocsar fishermen, if they find turtles in their nets, they butcher all of them, immatures, females, males, every last one. We would never take more than one or two adults from the same place—that's sufficient for all of us to eat."

I had never seen him so sternly distinguish his hunting practices from those of others. I wondered if his mood was triggered by the image of

Comcáac children playing with a baby sea turtle that was nearly their own age, their own size.

"We have this belief about sea turtles and certain other creatures. That's why it's prohibited to put a young tortoise or coyote in your own house. It's why I won't molest a nest of hatchlings, of ospreys or owls. Anything. You'll destroy the family, and bad luck will fall on your own family."

IN THE AUTUMN of 1994 I was invited to Hermosillo, Sonora, for a workshop that would supposedly decide the fate of protection for Islas Tiburón and San Esteban. On the flight down, I wondered what the term "protected area" actually meant. Protected from whom? From the Comcáac themselves? From outsiders? From all of humanity? Depending on which Mexican president's decree you read, these islands could be regarded as a National Wildlife Refuge (since 1963), as an indigenous reserve of the Comcáac (since 1975), as part of a special biosphere reserve that includes many islands (since 1989), as a proposed state protected area, or as a military reserve. Throw in pressures from private hunting guides, cruise ships, drug traffickers, or poachers, and it was clear that authority over these islands was up for grabs. It would take an unprecedented desire for consensus among many interest groups to arrive at some straightforward, inviolable agreement concerning the destiny of these islands.

I entered the conference hall early enough to sit back and watch as government officials, TV cameramen, newspaper photographers, conservation biologists, and ecotourist guides assembled for the "workshop"—a euphemism for a press conference to announce what had already been decided. Just before the entire assembly rose for a welcome and the reading of a declaration from the Sonoran governor's office, four Seri men joined the dignitaries at the front table. It was at that point that I sensed something interesting was about to happen.

To the surprise of most of us, several agencies in the federal and state governments were jointly proposing that the Comcáac community assume complete management authority over the natural resources of the islands. The proposal still required the approval of the Instituto Nacional de Ecología—and we didn't yet know that such approval would be withheld. We were simply told how happy everyone was with the draft management plan that had been prepared by Centro Ecológico staff, who in turn had consulted with a few key Seri leaders. (The Centro Ecológico de Sonora was a state natural history survey institute that has more recently been integrated into a state land use agency, IMADES.)

"Perhaps the area where our sacred sites are located on Isla Tiburón ought to be [what biosphere reserve planners designate as] the core protected areas. Not everyone can or should go there. Years ago, the father of Chapo [Barnet] was one of the few who was allowed to visit those places. It takes a special mental capacity to go to some of those places, because of their spiritual power. Not everyone can endure it."

ERNESTO MOLINA, *Punta Chueca*

If the plan were approved, we were told, the Comcáac community would have authority to implement it, and would receive enough federal support to equip their men to patrol the islands to stop illegal resource extraction. They would be encouraged to involve wildlife scientists and resource management technicians from the Centro Ecológico or from any other institution they selected to help them achieve the plan's objectives. It would mark the first time in history that scientists seeking to conduct research on the islands would be required to obtain a permit from the Comcáac and work under their discretion and direction.

At one point, an ethnobiologist representing the Instituto Nacional Indigenista in Mexico City officially invited the Comcáac to join the recently established Network of Indigenous Inhabitants of Biosphere Reserves, to learn how other Indian groups from Canada, the United States, and Mexico were handling "their biologists."

The most remarkable moment of the press conference, however, was when Tribal Governor Pedro Romero took over the microphone. He reminded the entire congregation that his people had been the legitimate managers and stewards of the two islands all along. In an eloquent philosophical tone, he posed a series of rhetorical questions to the audience: Why, he asked, were there so many animals still on these two islands that scientists could not find anywhere else? Perhaps because the Comcáac did not deplete their populations as some of their neighbors had done elsewhere. Why were animals like the Desert Tortoise more abundant in Comcáac territory than nearly anywhere else scientists had studied them? Perhaps because Comcáac spiritual traditions taught respect for these animals, and so they had reasons to refrain from hunting tortoises at certain times, or under certain conditions. Why should the Comcáac be the managers of this land and not the state or the federal government? Perhaps because the Comcáac had detailed local knowledge of the resources unique to the area, and had shared this knowledge down through the generations over centuries. Their ancestors had lived and died there. For that reason alone, it would be hard for them to ever let this land be destroyed or developed.

"Islas Tiburón and San Esteban and the midriff of the Sea of Cortés have great cultural and economic significance for us, the Comcáac. It is from here that our people obtained an important part of our nutritional and spiritual sustenance, and received the strength to survive times of great social or natural adversity."
PEDRO ROMERO, *tribal governor, speaking at the Centro Ecológico de Sonora, 1994*

At first glance, Pedro Romero's speech that day seems to echo many given by Native American leaders since the 1970s. It placed his people's knowledge of marine and desert wildlife in a position equal to, if not superior to, that of the academically trained resource managers who had been given authority over the region's wildlife in the recent past. Superficially, Governor Romero's speech seemed to reassert the claim that any Native American group must be considered "the original ecologists" of their homelands. Certainly, this strategy might give the Comcáac additional legitimacy as they sought to outcompete scientists from government agencies and become the official resource managers of the islands. While Romero welcomed "capacity building" and collaborations with technical consultants, in his mind these efforts should build upon and not replace the traditional ecological knowledge still held within the Comcáac community, a knowledge that has let them live in balance.

Assertions that Native Americans were "ecologically noble savages" have created controversy not only within the scholarly community but within indigenous communities as well. Did all Native Americans "live lightly on the land" in the past, and must they give up all modern hunting technologies in order to reaffirm their role as stewards of their homelands? Must the Comcáac, for example, demonstrate that their hunting with rifles has never depleted an animal population in order to be granted the authority to manage wildlife on Isla Tiburón?

What if a Seri individual has killed hundreds of rattlesnakes (as Roberto Camposano has) or eaten sea turtle eggs (as Carolina Hoeffer has)? Does that then preclude the possibility that all Comcáac are by nature, by gene, or by culture "ecologically noble"? Or does it suggest, as Henry David Thoreau concluded from his visit to Indian country in Maine, that once-noble native ecologists, in adopting the white man's ways, have "fallen" from their former nobility? And what, in turn, does that say about all those who are subsumed under the category of Cocsar, Yori or White Man? That they, by nature, can never be ecologically noble?

As environmental ethicist Douglas Buege (1996) has argued, "essentializing" all indigenous peoples as ecologically noble has all the ugliness of any racist stereotype, even when the stereotype is meant to be laudatory. Instead of buying into the stereotype of ecological nobility by birthright, Buege asks us to consider whether indigenous communities cannot be more tangibly regarded in terms of what he calls "environmentally epistemic responsibility." By that he means "a special type of authority that people gain through their immediate life experiences; they are experts concerning those particular experiences. Native peoples often have important epistemic privileges concerning the particular environments that they inhabit—e.g., they may have detailed knowledge of ecological relationships, species types, or particular ecological roles. They also frequently know the consequences of certain human actions upon their environment." This notion is different from granting the status of ecological nobility by birthright: "Claims of knowledge gained through epistemic knowledge . . . are confirmable because they involve particular observable aspects of a people's relationship to the land; they are not offered categorically and do not suggest that there is some essence which all people in the group share. . . . Relationships of native peoples to land are quite complex and need to be understood (and respected) in order to determine whether or not those people are capable stewards."

Pedro Romero was not asking that his fellow community members be granted Ph.D.'s in ecology on the basis of their DNA. Rather, he, much like Buege, was asking us to pay attention to traditional Comcáac practices that tangibly benefit wildlife. If we accept this invitation, we must consider that the Comcáac—or any other cultural community—may also have practices that adversely affect certain wildlife species.

One of the conundrums in addressing such issues lies in defining what a "traditional practice" is. For instance, does bow-and-arrow hunting qualify, but not hunting with rifles? Tropical wildlife ecologists Hames (1991) and Redford (1989) have demonstrated that a change in weaponry alone makes certain wildlife more (or less) vulnerable to overhunting, even when the hunters are still defined as "traditional people." Did outboard motors and nets increase the predatory efficiency of sea turtle hunting over that which harpooning from balsas allowed, making these animals more vulnerable to local depletion? Yes, definitely. But is that why most sea turtle species are scarce in the Sea of Cortés, and several are globally threatened with extinction? Probably not.

"In the early 1960s, when Ike Russell first took me to the Seri region, there were still lots of sea turtles. . . . Originally when the Seri interacted with sea turtles and other wild resources, it was on a subsistence basis, and they probably had very little

Figure 36. A Comcáac woman preparing to butcher sea turtles following a hunt on Isla Tiburón. Note "Little Bull" at the right using a rock to imitate the photographer's camera and picture-making activities. (Photo #6 by William N. Smith, 1946–51, courtesy of University of Arizona Special Collections)

impact on the resources. Beginning with the roads put through in the 1930s and 1940s, and escalating after World War II, the men started commercially fishing and hunting sea turtles. It is said that in the old days they never bothered with small turtles and only brought in big ones. . . . As sea turtles became scarce and prices escalated, a turtle of any size was a prized commercial commodity. There is a mania in Mexico for sea turtle meat. For one thing, the meat is delicious, and for another thing, there is the belief that it is good for virility. Tired old men are willing to pay high prices for sea turtle meat."

RICHARD FELGER, *Tucson, Arizona, 1990*

Even with the population recovery that the Comcáac community enjoyed this century, it is unlikely that they ever harpooned more than two to three thousand sea turtles per year (mostly the more abundant Green Sea Turtle, *Chelonia mydas*), even at their peak as commercial turtle hunters (Fig. 36). In the last decade, however, sea turtle harvests by the Comcáac have been more on the order of only twenty to fifty individuals per year. Most of those intentionally hunted were Green Sea Turtles for ceremonial purposes, although a few Hawksbills were also rescued from fishermen's nets. I have not personally witnessed any commercial sales of sea turtle meat by the Comcáac in recent years.

In contrast, I regularly hear rumors that non-Indian fishermen in Guaymas and Kino continue to sell sea turtles through black markets in Hermosillo and other northern Mexican cities. As late as 1990, Kim Cliffton observed freezer trucks full of the meat of Leatherback and Loggerhead turtles captured off Baja California coming in to Sonoran restaurants through the port at Guaymas. Farther south in Mexico, Cliffton (1990) heard of Leatherback turtles still being offered in restaurants, and long nets being set offshore from nesting beaches to obtain the hides of Leatherbacks as well as meat for shark bait. In the case of the Leatherback, global pressures have so pervasively depleted populations that these turtles almost never arrive in Comcáac village waters anymore. The only Leatherback I have seen over a quarter century of visiting the Sea of Cortés was in 1976, when Kim Cliffton, Richard Felger, and I came upon one dead on the beach near El Golfo de Santa Clara, less than 30 kilometers south of the Colorado River delta (see Felger 1990). In 1980, sea turtle biologists estimated that thirty thousand female Leatherbacks nested between Baja California and the Isthmus of Tehuantepec. By 1996, less than nine hundred females were reported using the customary nesting beaches along the entire Pacific coast of Latin America. This led James Spotila et al. (1996) to offer the following prognosis:

Leatherbacks are on the road to extinction and further population declines can be expected unless we take action to reduce adult mortality and increase the survival of eggs and hatchlings. . . . As a maximum, there should be a limit of 1% on the total annual mortality allowed to occur in any adult Leatherback population due to all human activity. . . . Even subsistence harvesting by indigenous fishermen on nesting beaches and in offshore waters is no longer sustainable by Leatherback populations. A level of harvest that was sustainable when human populations were low is no longer sustainable when human populations have expanded greatly near nesting beaches and foraging grounds of Leatherbacks. The same proportionate harvest per person is now a devastating level of exploitation.

These turtle conservationists are probably correct to warn that if Leatherback populations are ever to recover, there should be no excuse for any killing of these endangered sea turtles by anyone. And yet, I worry that by being so adamant as not to allow concessions to indigenous fishermen, the effort to protect Leatherbacks remains unfocused on the most critical threats. Globally, shrimp trawlers continue to kill an estimated 150,000 sea turtles a year and treat Leatherbacks with no more deference than they do other turtle species. While Mexico has agreed along with other countries to mandate turtle-excluder devices (TEDs) on all trawlers operating within its waters, trawler operators resent this imposition. Neither the Mexican wildlife wardens in the PROFEPA agency nor the Mexican navy perform enough open-water vigilance to know whether the devices are regularly used.

Three marine biologists have independently reported trawlers leaving ports in the Sea of Cortés with TEDs on their nets, then switching to other nets that capture turtles, which are sold through black markets to elite restaurants. Drift nets, plastic wrappers (which look much like the turtles' prey, jellyfish), and the poaching of eggs from beaches also pose threats. Yet the damage caused by trawling may be an order of magnitude more severe, for not only does it kill or injure turtles directly, but it also degrades their resting grounds and eliminates their food sources.

And yet, there is some hope for sea turtles in the Sea of Cortés. In January 1999, Environmental Flying Services pilot Sandy Lanham observed evidence that as many as fifty Leatherbacks might still be coming onto beaches as far north as Baja California to lay their eggs. Some Seri individuals are currently planning to visit nesting beaches in Central America to internationally express their concern that egg poaching and trawler harvests continue to imperil their most important ceremony, the fiesta for *xica cmotómanoj*, "the fragile ones" (Figs. 37 and 38). Gabriel Hoeffer, a young Seri conservationist, went on national television with the Grupo de Cien to urge Mexican policymakers to provide stronger protection for sea turtles in the new millennium, and has since coauthored a paper on his recapture of a tagged Green Sea Turtle (Seminoff et al. 2002).

"Yes, we would like to [go to Central America, to] see where Leatherbacks nest and to help those places be protected. But it is not a thing we could take lightly; any Seri who went would have to be prepared to help offer the entire Leatherback ceremony to all of those present, painting their faces, singing, everything."
IGNACIO "NACHO" BARNET, *tribal governor*

Conservationists often call upon cultural communities to practice restraint in the face of overexploitation of traditional resources, regardless of whether the communities' own actions have been a primary cause of depletion. An ethic of restraint is already practiced in Comcáac villages. Young Seri conservationists regularly confront their neighbors who are found attempting to catch sea turtles trapped in their fishing nets instead of releasing them. The Leatherback is sacred to the Comcáac, revered in a four-day ceremony that is performed whenever they find a beached animal. In the ceremony, the turtle is always released into the water at the end of the four days. The survival rate of these animals has apparently been quite high, judging from oral histories recalling the last six decades of Leatherback ceremonies. Moreover, the Comcáac never seek out Leatherbacks for eating or for sale to others; the very thought of doing so is repugnant to them.

Similarly, Comcáac artisans willingly switched to carving in stone and soft wood rather than ironwood when their non-Indian neighbors depleted this tree for charcoal production and machine-made tourist crafts—even though, according to my own field measurements, Comcáac harvesting practices appear to be sustainable.

Figure 37 (*right*). A
Leatherback Turtle being
painted and blessed with
sacred herbs. (Photo #2 by
William N. Smith, early
1960s, courtesy of
University of Arizona
Special Collections)

Figure 38 (*below*). A live
Leatherback Turtle painted
for a ceremony. (Photo #4
by William N. Smith, early
1960s, courtesy of
University of Arizona
Special Collections)

The Comcáac have been affected by another conservation debate as well, one that questions whether indigenous hunters are just as predatory as nonindigenous peoples. Kent Redford (1989), for example, argues that indigenous people have been no less prone to overhunting than others have, but that many game populations on native hunting grounds survived historically because these hunters lived in lower densities, and had less access to powerful weaponry, than present-day hunters do. While it is true that historic Comcáac settlement densities were extremely low except along the coastlines, they were high enough to pose threats to endemic animals such as the San Esteban Chuckwalla and the Desert Tortoise. If these reptiles were hunted daily or weekly by even a few family groups, a few thousand chuckwallas could easily be depleted from a island such as Alcatraz or San Esteban in less than a decade. And indeed, they've got the reputation for it. As Steinbeck and Ricketts observed in their book *The Sea of Cortez* (1941), the Comcáac have been denigrated by other cultures as being ruthlessly predatory.

Zooarcheologists Storrs Olson and Harry James (1984) have demonstrated that flightless rails and crakes disappeared from island after island in Polynesia, hunted out by prehistoric seafarers who foraged in groups no larger and no better equipped than the prehistoric Comcáac. Their colleague Jim Mead has begun to explore whether bone deposits recovered from caves on the islands of the Sea of Cortés include any evidence of local extinctions of flightless birds or reptiles there as well. To date no such evidence has been found in Comcáac hunting grounds, although the probability of at least some historic island extirpations of vertebrates remains high.

"The people of Isla San Esteban once managed the chuckwallas there when they became scarce. I think they are becoming very scarce once again, due to all the rats escaping from boats and eating their eggs."
ERNESTO MOLINA, *Punta Chueca*

What we do know is tenuous, yet suggestive: certain taboos may have reduced the frequency of consumption of certain reptiles by the Comcáac. Without these taboos, these animal populations could have been depleted as rapidly as Green Sea Turtles were in the Sea of Cortés during the free-market competition of the 1970s. Descendants of those who once lived on Isla San Esteban, for example, said that such taboos were in effect on that island and were rooted in pervasive spiritual beliefs. From the beginning, Hast Cmique instructed them that it was "wrong to kill animals without reason, such as wounding them and leaving them to die. A person who mistreats a chuckwalla by throwing it to sea will be punished when he is at sea, perhaps by being subjected to a strong wind" (Bowen 2000).

"There is another thing they say about going out hunting, for Mule Deer, anything. If they don't respect the Desert Tortoises when they are out there, they will suffer a sickness. They used to hunt a lot [on Isla Tiburón], but they had to leave the tortoises alone while they were out hunting, or they would have bad luck with what they were doing. . . . Desert Tortoises have a spirit different from that of other animals."

JESÚS ROJO, *Punta Chueca*

Other taboos may have worked to reduce the vulnerability of Desert Tortoises—which have a long history of interacting with the Comcáac and, indeed, are believed to understand Cmique Iitom—to overhunting. Specifically, hunters going out to kill Mule Deer or other big game were forbidden to capture and kill any tortoises they encountered along the way. It is said that the ease of capturing the tortoises would break the concentration a hunter needed for stalking and shooting a deer. To kill a tortoise under such circumstances would bring a Seri individual bad hunting luck, not merely on that foray but on future ones as well (Nabhan 2002). Some Seri, however, are very cautious about offending tortoises under any circumstances, believing that it will bring physical and spiritual harm.

In addition, young women are strongly discouraged from eating Desert Tortoise meat. If a woman gives birth only to girls, people explain that she must have eaten the reproductive organs of female tortoises when she was pregnant. If she gives birth only to boys, they explain that when she was a girl, one of her friends must have thrown a tortoise penis at the small of her back.

However strange these beliefs sound to outsiders, they certainly put a damper on a Seri hunter's willingness to hunt tortoises, or a Seri woman's willingness to eat tortoise meat or even be present at the butchering.

"Desert Tortoises can speak and they can be spoken to. Ziix Taaj was the only person in the world who knew what they were saying. He would call up their form—like a mirage—and they would speak in front of him while they were gambling. Perhaps that is why it is prohibited to leave a baby Desert Tortoise locked up in your house, where no one of its kind could ever grow. The mother or father of that tortoise will suffer terribly for lack of their little one."

ALFREDO LÓPEZ, *Punta Chueca*

Do such taboos always function to discourage overhunting? Probably not, or if so, to only a minor extent. Ethnobiologists have a tendency to read ecological messages into taboos such as these, and few indigenous people would ever baldly explain them as a cryptic cultural means to reduce resource overexploitation. And yet, such taboos seldom exist in an ecological vacuum; they may work to reduce pressures on resources under certain conditions, regardless of whether that outcome is understood by all practitioners.

"There used to be a lot of Desert Tortoises over at Campo Coralito, but they have suffered a lot of losses due to their capture by fishermen boating over there from Bahía Kino. They sell them for turtle soup."

ERNESTO MOLINA, *Punta Chueca*

Even if reptile populations were historically kept from depletion by these taboos and by low human population densities, many additional pressures have been exerted on these populations over the last half century (Hames 1991). Today, all marine reptiles are suffering from the effects of shrimp trawling, diminished nutrient flow from the Colorado River, and contamination in the estuaries edging the Sea of Cortés. Migratory sea turtles remain vulnerable to trawlers, long-lines, drift nets, and entanglement in debris throughout their foraging ranges in the Pacific. Desert reptiles, though not affected as dramatically, are locally imperiled by car traffic, water impoundments, poaching, and altered wildfire regimes fueled by invasive buffelgrass.

While the Comcáac community cannot keep these pressures from impacting the animal populations in the homelands and territorial waters, they can make a difference in two other ways. First, they can take better care of the creatures on their *ejido* lands and in the Canal del Infiernillo, as twelve Seri men are doing in their newly gained capacity as "para-ecologists" trained by Northern Ari-

Figure 39. The initiation of the first class of Seri para-ecologist trainees by tribal elder Genaro Herrera, Isla Tiburón. (Photo Karen Krebbs, 1998)

zona University, Arizona-Sonora Desert Museum, Comunidad y Biodiversidad, and the Amazon Conservation Team (Figs. 39 and 40).

In 1998, the Comcáac collaborated with Mexican government agencies to erect signs on the edges of their territory and at fish camps on their islands, warning visitors not to collect or damage plant and animal life. One of the para-ecologists has even chastised his cousins for eating sea turtles found in drift nets, rather than releasing them.

"Yes, we bear arms [to protect our territory]. Isla Tiburón belongs to the Comcáac community, and those [outsiders] who fish around it are thieves. If the government authorities don't help us, then the community will provide its own vigilance."

ANONYMOUS, *Punta Chueca*

Figure 40. The initial gathering of the para-ecologists for instruction in Punta Chueca. (Photo Karen Krebbs, 1998)

The Seri thus have been able to punish poachers, clandestine trawlers, and live-animal collectors. They have also been bold in confiscating pangas of fishermen and turtle hunters who venture into their waters without permission, as is attested by a *corrido* written and sung by the acoustic troubadours of Bahía Kino, Los Ribereños del Golfo. In their ballad of a dispute between the Comcáac and neighboring, non-Indian fishermen, "El Canal del Infiernillo," listeners are warned "not to get stuck in the Canal," where the Comcáac have control, for the price is high: a million pesos ransom for the return of boats and gear. In an incident during the winter of 1996–97, young Seri activists boarded a shrimp trawler armed with automatic weapons after its captain refused to move from tribal waters, telling them that he had paid "a fee" to one tribal leader, which gave him the right to trawl anywhere he wished in the Canal del Infiernillo. Although the activists were temporarily jailed, they were later acquitted, for they were legitimately defending their people's right to protect resources against interlopers. Finally, in the winter of 1997–98, three Mexican government agencies agreed in writing that the Comcáac tribal governor had the authority to enlist military support to expel such intruders should they refuse to depart from tribal waters.

Winning political battles such as these does not yet mean that tribal waters are free from trawlers, nets, or long-lines owned by non-Indian competitors. In September 1998, when I accompanied Alfredo López to Isla San Esteban, we witnessed numerous pangas and trawlers working near Ensenada del Perro on Isla Tiburón, as well as several sets of *changos*, or "monkeys," a kind of small-scale trawling apparatus whose use is officially prohibited in Mexican national waters (Cudney-Bueno and Turk-Boyer 1998). Obviously, someone had been "monkeying around" in Comcáac territory without any permit to do so, so Alfredo confiscated the equipment and turned it in to tribal authorities. In his exchanges with the non-Indian fishermen, however, he was cordial, asking them who they worked for rather than demanding that they immediately leave. Some Seri, of course, fear retaliation if they demand that the non-Indian fishermen obey the law and depart Comcáac territory.

"I am not angry with these poor young kids who are illegally fishing in our waters. It is the *patrón*, who grubstakes several boatloads of men, who tells them all where to fish. If the *patrón* is giving them orders to come into our territory, then we should go and talk with him. He is the offender."

ALFREDO LÓPEZ, *Punta Chueca*

The second means the Comcáac have for protecting their resources from external pressures is collaboration with organizations seeking conservation policy reform. By periodically participating in alliances with biologists in SEMARNAP (Secretariat of the Environment, National Resources, and Fisheries), CONABIO (Biodiversity Commission), IMADES (Institute of Environment and Development for the State of Sonora), Comunidad y Biodiversidad (Community and Biodiversity), Conservación Internacional (Conservation International), the Alianza Pro–Palo Fierro (Pro-Ironwood Alliance), the Amazon Conservation Team, and other such organizations, they have helped determine which species should be officially protected and how such protection should be implemented. Comcáac support was key to obtaining protection status for sea turtles and ironwood, for example, and now they are allying themselves with those calling for stricter control of shrimp trawlers in the Sea of Cortés.

The third means the Comcáac have recently used to establish their role as conservationists is to field-monitor sea turtle and Desert Tortoise populations and present their findings at international meetings. Recently, two Seri para-ecologists presented their own quantitative data on sea turtle size classes (Fig. 41) and coauthored a published paper on the migration of Green Sea Turtles from Michoacán (Seminoff et al. 2002).

While I have absorbed dozens of such papers on wildlife conservation in the Sea of Cortés region over the quarter century I have done research there, I am nevertheless constantly humbled by what Seri friends teach me about the animals of their homelands. Theirs is knowledge that will surely aid conservation efforts. Whether we choose to call such traditional knowledge "indigenous sci-

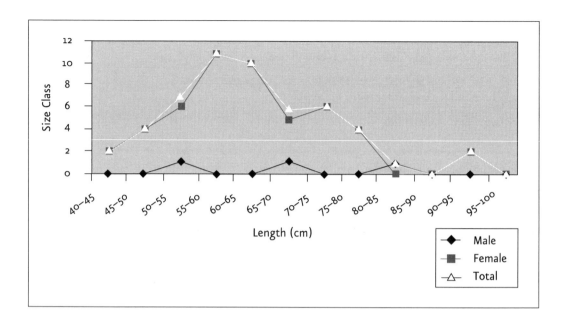

Figure 41. The size-class distribution of the carapaces of Green Sea Turtles from the Canal del Infiernillo rescued and measured by para-ecologists, 2000. (Data from Humberto Morales, Jeff Seminoff, and Gary Nabhan)

ence" or "folklore" or "sacred ecology"—or whether we choose to put no artificial label on it at all—there can be no denying that this orally transmitted knowledge can enrich our own vision of the natural world and inspire all of us to better protect it (Berkes 1999). I was never more profoundly aware of this fact than when Jesús Rojo valiantly attempted to interpret for me the significance of a brief but lovely-sounding song from the Leatherback ceremony:

"Well," he said, groping for words, "there's a story with this song, one that I don't know well. It was from a harpooning elsewhere. . . . See, they were out hunting [turtle], no? They weren't aboard a panga, for in those times they didn't use pangas, only kayaklike balsas of giant canes, but they too were plenty big, for they could hold three, maybe four persons, no?

"Well, when they harpooned one [of the turtles], they were far from camp, but that was when the Leatherback came along with them, no? It went into their camp on its own volition, right to where they beached their balsa boat. The Leatherback came in on its own, for it understood the words, the dialect of the Seri. And for that it was coming along, coming along until it reached the camp and arrived at the edge of the water.

"The sea turtle came rising up, it left the water, no? It reached toward where it was dry, and when it reached the beach, it *danced and rolled over*, this turtle did, it *turned over like this* [motioning], its carapace going under until its heart in its breast faced upward. *This is what they were singing for.*"

the historic decline and recent revival of traditional ecological knowledge

The deleterious effects of language loss and species loss are not merely additive but negatively synergistic. . . . [And yet, the vestiges of] indigenous knowledge systems may have even greater impact in informing Western conservation measures [by] establishing a deep conservation ethic based on . . . a deep sense of stewardship rather than ownership . . . of the natural world.

PAUL ALAN COX, "Indigenous Peoples and Conservation" (1997)

Alfredo López reminded me that an elderly woman named Isabel Torres
had asked that I sing her the Regal Horned Lizard song one last time before
I departed for Arizona. He wished me well and directed me to the shade
tree where he had last seen Isabel sitting. I found her easily, and as I
greeted her, she turned in the direction of my voice; I then recalled that
she was blind. I sat down in the sand next to her and held her hand so
that she could be sure who and where I was.

As I sang the *hant coáaxoj* song to her, she laughed sweetly. When I
finished, she touched my shoulder.

"Come a little closer," she told me. "My voice is not strong because I am
an old lady, but I have something I want you to know."

I moved in close to her, my ear next to her mouth.

"You haven't been told, have you? That the Regal Horned Lizard, the one
you sang of in the song, well, he was a person. You thought all this time
that you were merely singing about a little animal, didn't you? Well, it was
a long time ago, but this man on Isla Tiburón went looking for firewood for
us, for his people. But he didn't ever come back."

I thought of the words of the song in this new light. "The ants?" I asked. "The ants killed him?"

"Yes, the ants, the kind that Regal Horned Lizards eat."

"Ahh." I began to understand. "And so they found only his remains . . . and the firewood."

"They knew it was him by his bones. That happened a long time ago, of course, in the time when that man lived whose name is the same as the little animal's. I just wanted you to know, because we've given you and one of the little boys around here the very same name. Hant Coáaxoj."

She let go of my hand and chuckled again. "And all this time you thought you were only singing about animals."

My last stop was to visit the retired sea turtle hunter Ramón Perales.

"Hant Coáaxoj!" he greeted me.

I then sang the Regal Horned Lizard song for him. And then, just in case I hadn't correctly understood Isabel, I framed a question for him in Spanish: "El camaleón que está en la canción, pues, dígame, ¿era gente una vez?"

Ramón laughed. "Of course! He was a man before the last world was destroyed and this world was created. He died when he went out to get wood for our people of those times. They later found his bones. But his life went on, and when this new world was created, his life came into this little animal."

"The man from those ancient times? Do you mean—how do you say it? His spirit . . . in the animal?

Ramón nodded. "Yes, Hant Coáaxoj. What you would call his soul."

WHILE MANY SERI individuals would express remorse over the senseless killing of Regal Horned Lizards in their world (Figs. 42 and 43), there is another kind of loss that they speak of with just as much emotion. It is the loss, particularly among the young people, of the traditional ecological knowledge that has served as the rudder for their culture's journeys into the natural world over many centuries.

Figure 42 (*left*). A Regal Horned Lizard carved from elephant tree wood. (Photo Helga Teiwes, 1997)

Figure 43 (*below*). A Regal Horned Lizard design in a miniature basket by Patricia Moreno. (Photo David Burckhalter, 1999)

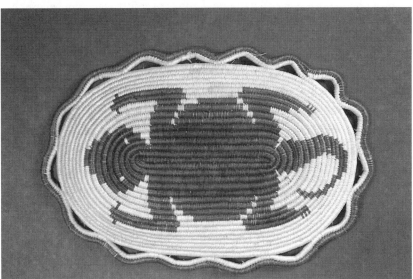

"Today, they [the youth] don't go to the island. . . . When I was a boy, my family would camp out there alone, away from the others. They'd eat wild honey, venison, tortoise every now and then, and plenty of sea turtle. My father showed me how to hunt. But now, those days are gone. It's been so long that I myself have almost forgotten the taste of sea turtle meat."

JESÚS ROJO, *Punta Chueca*

Even if currently threatened species recover in time, the Comcáac are not so sure that their own traditions associated with those plants and animals will. They realize that young people are exposed to Spanish as much as to their own native language, and that fewer and fewer families depend on locally harvested resources for their diet (Fig. 44). Few young couples, elders speculate, could bring in enough food to survive if stranded on Isla Tiburón or at a remote location along the Sonoran coast. Comcáac adults over fifty generally believe that few youths now learn orally transmitted ecological knowledge in enough detail to maintain Comcáac traditions over the long haul.

Is this belief a reasonable one? While endangered species and endangered languages have been well studied, only a handful of empirical studies have been published assessing the degree to which traditional ecological knowledge is being lost generation by generation. These pioneering publications—most notably Ohmagari and Berkes 1997, Zent 1999, Zitnow 1990, and Rosenberg 1997—all point to one key fact: much of this linguistically encoded knowledge is transmitted to younger generations while they are assisting elders in traditional hunting, fishing, or foraging activities, or while food, fibers, medicines, or ceremonial items are being prepared by hand (Nabhan 1998a).

Immersion in such traditional subsistence activities is also essential to the very transmission of native languages. The conclusion of John Edwards's 1995 study of multilingualism bears repeating: "To put it simply: a decline in the existence and attractions of traditional life styles also inexorably entails a decline in the languages associated with them." Just to the north of Seriland, Tristan Reader (1997) of Tohono O'odham Community Action has articulated what many of

Figure 44. A young girl from Desemboque accompanies her aunts, who gather medicinal plants in the wild. (Photo Christine Keith, 1997)

his Native American co-workers have been feeling for some time: "It is not enough to preserve a language, its words and its linguistic structures, its images and grammar. In order for a language to be truly alive and vital, we must also preserve the subjects of discussion . . . that grow out of ways that people make their living. . . . Cultural preservation requires that we rejuvenate traditional food systems, local economies and ways of interacting with the natural world."

In February and March of 1998 I attempted to assess whether the younger members of the Seri community are still engaged enough in reptile-related traditional activities to ensure that Comcáac cultural values and specialized vocabularies will continue to be passed on (Nabhan 1998a). Instead of focusing on vocabulary or on stories and songs as indicators of cultural persistence, as Rosenberg (1997) and Zent (1999) did, I chose to follow Ohmagari and Berkes (1997) and ask elders to identify key culture-transmitting rites, skills, and practices in which they had participated by the age of ten. I then devised a battery of ten questions having to do with various aspects of Comcáac engagement with reptiles (Fig. 45).

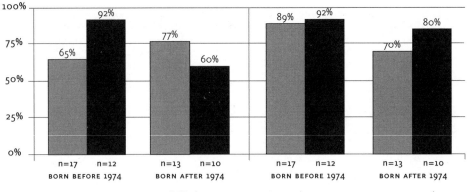

45a. Have you ever killed a
rattlesnake for its products?

45b. Have you ever captured
a live Desert Tortoise?

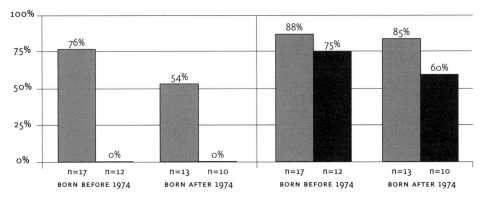

45c. Have you ever harpooned
a sea turtle?

45d. Have you ever cooked a
sea turtle in its carapace?

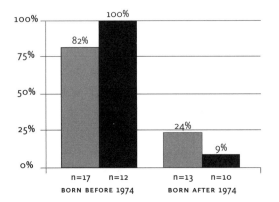

45e. Have you ever attended a
Leatherback ceremony?

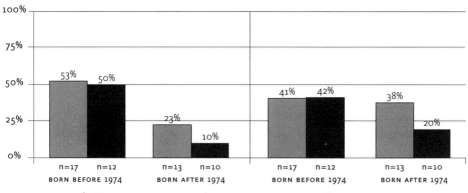

45f. Have you ever treated
snakebite with herbs?

45g. Have you ever eaten
Desert Tortoise eggs?

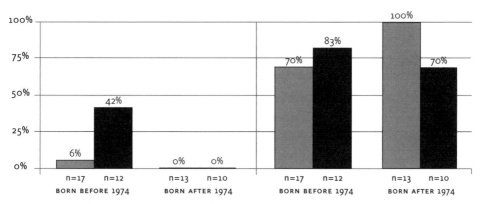

45h. Have you ever moved
a sandsnake over the waist of
a pregnant woman?

45i. Have you ever taken care of
a horned lizard (as a pet)?

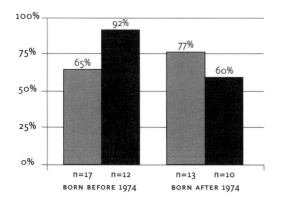

45j. Have you ever captured
a gecko?

Figure 45. Responses of Seri individuals
on their engagement with reptiles.

"The youth of today don't understand the language well, but what's more, they don't eat Desert Tortoise, venison, or sea turtle. When I was young—maybe ten, eleven years old—my master [father] took me to teach me how to hunt, harpoon, and gather water. I had to learn to carry twenty liters of water long distances. Then my father would sing to me many times the songs he wanted me to know. My mother also taught me songs. Now the kids may know how to use the *chinchorros* [nets] to fish, but they don't know how to harpoon. . . . They don't know the plants either."

JESÚS ROJO, *Punta Chueca*

I had no trouble obtaining one-on-one interviews with twenty-nine adults, most of them born well before 1974—which is about the time the Mexican government established an office to administer services to the Comcáac, including the building of permanent housing. I asked the questions in simplified Spanish, repeating key words or phrases in Cmique Iitom, sometimes acting out certain activities to ensure full understanding of the question. I recorded affirmative and negative answers to the ten questions and took notes on respondents' commentaries about these and related activities. (These commentaries contributed to the "species accounts" presented in part 2.)

I also attempted to obtain one-on-one interviews with younger Seri individuals, born between 1974 and March 1988 (so aged ten to twenty-four). In the end, however, I had to modify this plan slightly. Often the older sister, brother, or mother of a youth remained present for the interview and would discuss the questions with the interviewee in Cmique Iitom. While I was at first concerned that the interviewee might be swayed by the views of older relatives and offer more "politically correct" answers, I gradually came to realize that having the questions clarified in this manner was useful. Even in the presence of family members, nearly all the Seri youths were unabashed in admitting which traditions they had yet to experience. They also became more enthusiastic in relating a past encounter with a reptile, and family members encouraged them to include details to fill out their accounts.

All in all I obtained twenty-three interviews with these youths to complement the twenty-nine interviews with their elders. The gender balance was 58 percent male to 42 percent female (as opposed to the 48 percent male/52 percent female balance officially recorded for Desemboque and Punta Chueca residents of Comcáac descent in 1995). My sample included about 10 percent of all Comcáac family members in the two villages. (About eighty-four families call Punta Chueca home, and seventy-five call Desemboque home, although these statistics are constantly in flux.)

Figure 45a illustrates one of the most interesting role reversals I uncovered during my interviews. While 92 percent of all adult women interviewees claimed to have killed a rattlesnake for its products—skin, vertebrae, rattle, or meat—only 60 percent of the girls had done so. At the same time, the percentage of boys who answered in the affirmative (77 percent) was 12 percent higher than that of the men.

Perhaps so few girls are now inclined to kill rattlers because of cross-cultural peer pressure: they know that neighboring Mexican girls would be unlikely to do such a thing. The boys, however, remain active in bringing home rattlesnakes so their mothers and grandmothers can make necklaces out of the vertebrae and rattles. The men still kill rattlesnakes as well, but none claimed to make anything out of them. They often kill them to rid their fishing camps of danger, but that was outside the focus of the question I posed. In the case of rattlesnakes, because local accessibility to this resource has remained high, it would seem that levels of involvement with harvesting and use have occurred mainly for cultural reasons.

The responses shown in Figure 45b demonstrate a much more straightforward trend: when compared with adults of the same gender, 12 percent fewer girls and 19 percent fewer boys have captured Desert Tortoises live despite the local abundance of the animal. Even more revealing was the decline in using trained dogs to hunt Desert Tortoises, a traditional strategy that appears to have died out entirely in recent years, since only a few individuals, all of them over forty-five years of age, had ever gone tortoise hunting with dogs. Interestingly, one young adult and his son claimed that the Comcáac never hunted Desert Tortoises because it was against traditional law; apparently they were confusing certain case-specific taboos with a total ban on tortoise hunting, for numerous reports and photos demonstrate that the Comcáac have indeed hunted and eaten tortoises for many decades.

Figure 45c demonstrates a straightforward decline in male participation in a traditional subsistence activity, the harpooning of sea turtles. Only one mythic woman was known historically to have harpooned sea turtles, aided by her husband's ghost. Perhaps because of the global decline in sea turtles over the last few decades, 22 percent fewer teenagers and young men than older men have harpooned any species of sea turtle. However, a technological shift has also taken place among Comcáac fishermen: nowadays, fewer spend time harpooning, and more work to master the placement of drift nets and hooked lines. Boys once grew up spending much of the warmer months wading out into the Canal del Infiernillo to practice harpooning; today, few boys are seen mastering this skill.

As Figure 45d shows, when sea turtles are brought in by net or line, fewer young people of both genders participate in traditional roastings of the meat than in the past. The method involved building a fire on the plastron of an overturned turtle; the meat then roasted inside the carapace. Today, with stoves, elevated cookfires, metal cookware, and potable water, sea turtle meat is more frequently boiled in a pot or grilled. This downward trend in participation probably is a reflection more of technological change than of sea turtle scarcity.

A telling indicator of sea turtle declines can, however, be seen in the answers to the question "Have you ever attended a Leatherback ceremony?" Figure 45e charts the dramatic decline in participation in this rite, in which a live Leatherback is sung onto the beach, painted, cared for, then released. Although many of the youths I interviewed initially claimed that they had participated in the full ceremony, most of them had in fact only observed the 1997 rite in which the bones and carapace of a dead Leatherback encountered near Punta Chueca were used as an opportunity to teach younger people about aspects of the ceremony. With perhaps as few as four hundred nesting females remaining along the entire Pacific coast of Latin America, fewer and fewer opportunities have arisen for the Comcáac to even see this species, let alone offer it their blessings.

While exposure to venomous snakes is still much more probable in and near Comcáac communities than in more urbanized, non-Indian Sonoran communities nearby, Figure 45f suggests a dramatic decline in the use of herbs to treat snakebites. While roughly half of the older adults of both sexes had been involved in treating the bites of rattlesnakes or seasnakes with herbs, the same was true of only 23 percent of the boys and 10 percent of the girls. This may simply mean that snakebite treatment is put into the hands of the oldest person present at the incident, given the risk that an envenomated bite poses. However, when a group of Comcáac are working or traveling together, everyone present may be asked to play a supporting role in emergency medical care. The younger people who had been involved in such a scenario, it turned out, had either been bitten themselves or had assisted an older person managing the treatment. In some cases, youths had to go and gather the appropriate herbs while the elderly care manager stayed with the victim. Participation in snakebite treatment, therefore, is not entirely age-dependent, or at least was not so in the past.

I believe, however, that more rapid access to medical clinics is the main reason for the great decline in the treatment of snakebite with herbal remedies. Older men are the cohort in the Comcáac population that has the most exposure to rattlesnakes on some of the islands, and to seasnakes while fishing in remote areas where knowledge of traditional remedies can mean the difference between life and death. Not only are younger men less likely to stray as far in their families' boats as their predecessors did, but if bitten they are also more inclined to drive to clinics an hour away than to search for medicinal herbs on foot.

Figure 45g shows a much sharper decline in the consumption of Desert Tortoise eggs among girls than among boys relative to their elders. Eggs are typically encountered during foraging for wild foods or other materials; they are not intentionally sought out by many families today. Although it is plausible that fewer eggs are laid near the two permanent Comcáac villages than around the scatter of camps on Isla Tiburón or on the mainland, I would guess that consumption rates more directly reflect the time a Seri youth has available for leisurely foraging.

Figure 45h demonstrates the possible loss of a Comcáac custom relating to the lovely Bandless Sandsnake (*Chilomeniscus stramineus*), known to elders as *hapéquet camízj*, which was once used to bless a pregnant woman and her baby with its beauty. Elderly Seri women recalled being present when someone would capture this little snake by hand, then move it across the belly and small of the back of a pregnant relative, to make the forthcoming baby as sleek and unblemished as the snake itself. The younger generation, in contrast, hardly knew the name of this snake, let alone of its presence in their village. The older people who fondly recalled this custom confessed that it is hardly practiced anymore. When they explained the ritual to younger Seri individuals, they were met with bemused bewilderment.

"*Hapéquet camízj?* It lives under the sand. When a woman is pregnant, her husband catches it and walks it across her stomach. When the baby later comes out, oh, such a pretty face, lovely eyes, a well-shaped nose!"

MARÍA FÉLIX, *Desemboque*

Fortunately, some reptiles have not become more remote from contemporary Comcáac life, as Figure 45i verifies. Regal Horned Lizards are still kept as temporary pets by virtually all boys, and 70 percent of the girls. These endearing little creatures continue to be encountered frequently, and handled with care, even love. Although most Seri kids who find a horned lizard keep it for just a few hours, then put it back where they found it, one young man recalled having raised a horned lizard to sexual maturity.

Finally, curious changes are occurring in the interpretation of the Comcáac belief that one should not engage with geckoes (Fig. 45j), a belief discussed elsewhere in this book. While 92 percent of the older women claimed to have captured geckoes, most of them did so to rid their household of them, presumably so that they would not be psychosomatically affected by them. Compare that figure to the 60 percent of girls and young women who have captured geckoes. This difference may indicate that Comcáac girls are being swayed by the ophidiophobia of their Latin American neighbors, which may be overriding the Comcáac belief that geckoes should not be touched for supernatural reasons. I suspect that 12 percent more boys than older men have captured geckoes because the boys put less stock in the notion of supernatural powers. Depending on how one interprets the data presented in Figures 45a–j, the estimated rates of loss in reptile-related traditions within the Comcáac community are quite high: as much as 23–30 percent among women, and 8–25 percent among men. Are these declines due to decreased abundance of reptiles or to changes in human behavior and values? In Table 23 I present my speculations on this question. Although the declines in the numbers of sea turtles reaching the Canal del Infiernillo have certainly disrupted Comcáac traditions, no good data exist on the population stability of Desert Tortoises, rattlesnakes, and other reptiles. However, the demise of some traditions has surely been caused not so much by reduced availability of animals as by reductions in the amount of time youths have available for foraging and listening to traditional stories that encode cultural beliefs. Attending school, watching television and listening to the radio, and reading magazines usurp the time once dedicated to mastering certain subsistence traditions. A trend toward lessened contact between youths and elders since 1930 has been well documented by Zitnow (1990) in other indigenous communities.

"There are boys who can't even count to fifty in their own language. There are youths who hardly use [Cmique Iitom] when they are playing together, or who don't use it at all when they are outside of our villages. There are even ten- to fifteen-year-olds who can't or won't speak."

PEDRO ROMERO, *Punta Chueca*

TABLE 23

Summary of Interview Findings on Declines in Participation in Reptile-related Traditions

	Percent Change			Potential Causes for Decline			
TRADITION	MALE YOUTH	FEMALE YOUTH	ANIMAL SCARCITY	TECHNOLOGY SHIFT	VALUE SHIFT	LESS TIME OUTDOORS
Killing rattlesnakes for products	+12	−32	No	No	Yes	Yes[1]
Live-capturing Desert Tortoise	−19	−12	No?	No	Yes	Yes
Harpooning sea turtles	−22	0[2]	Yes	Yes	No	Yes
Cooking sea turtles in shell	−3	−15	Yes	Yes	No	No
Attending Leatherback ceremony	−58	−91	No	No	No	No
Treating snakebite with herbs	−30	−40	No	Yes	Yes	No
Eating eggs of Desert Tortoise	−3	−22	No?	No	Yes	Yes
Blessing pregnancy with sandsnake	−6	−42	No?	No	Yes	No
Caring for pet horned lizard	+30	−13	No	No	No	No
Capturing a gecko	+12	−32	No	No	Yes	No

Notes: This table charts the decline from Comcáac born prior to 1974 to those born after 1974. Increases (+) may or may not mean shift away from tradition.

[1]Females more strongly affected.

[2]Women have not traditionally participated in harpooning.

Is acculturation a factor? Increased exposure to Latina norms of social behavior, including ophidiophobia, has certainly disrupted Seri girls' involvement in traditional activities that they now consider unladylike, uncouth, or even dangerous. With regard to technological change, fewer Seri young men rely on medicinal herbal remedies than on medical clinics, and they know less about harpooning sea turtles or hunting tortoises with dogs because they are learning about nets, lines, cars, and motorboats.

These changes do not mean that the Comcáac have been fully assimilated into mainstream Mexican society. On the contrary, they continue to distinguish themselves from their non-Indian neighbors in numerous ways, ways that have effectively been used by other "persistent peoples" to stave off acculturation. Nevertheless, it appears that the loss of traditional ecological knowledge is proceeding, and Comcáac community elders and educators are increasingly concerned about the potential consequences of this loss.

For outside observers to grasp the consequences of this loss, we must ask questions that are relevant to us all, no matter what our ethnic background or place of residence: How is local knowledge of the natural world valuable, and for whom? In a world facing unprecedented environmental, economic, and social change, is it possible for some of this knowledge to become obsolete or diminished in value? Or do some components of traditional knowledge have a timeless quality, providing a legacy that will continue to enrich future generations? If so, how can we more fully honor that knowledge, including by sustaining the animal populations of which it speaks?

Because of their concern about traditional knowledge loss, Comcáac leaders have requested that their collaborators in conservation give as much attention to these issues as to the issue of species survival. Whenever I have discussed with Comcáac leaders how best to return the benefits of conservation biology research collaborations to their communities, we have agreed that any newly synthesized information be adapted for use in schools and town meetings. They want their youths to learn about the animals native to their homeland, and the traditions associated with them. This information sharing continues.

Since these concerns were first communicated to me by the late Tribal Governor Pedro Romero, I have worked with a number of talented professionals in what has been loosely known as the Seri Ethnozoology Education Project. Supported by the Amazon Conservation Team, Arizona-Sonora Desert Museum, the Columbus (Ohio) Zoo, and Northern Arizona University and by the donation of many hours and many incidentals from individuals within and beyond the Comcáac community, this project has worked hard to integrate indigenous science and Western science in ways that benefit local communities.

The first phase began with Pedro Romero's suggestion that my interviews with elders about their knowledge of reptiles be published as a booklet that Comcáac children could use as part of their schoolwork. With guidance from Pedro, herpetologist Howard Lawler, and myself, cross-cultural educator Janice Rosenberg devised a primer using the Comcáac genre of teaching through riddles, nursery rhymes, and pictures, entitled *Animalitos del desierto y del mar* (1997; Fig. 46). Distributed to all Seri classrooms in 1997, the booklet was embraced enthusiastically by Comcáac children. Rosenberg later went on to edit a second bilingual edition, which included drawings and stories recorded by the children themselves; that too has been distributed to the primary schools in Punta Chueca and Desemboque.

"The Seri try to protect the wildlife more than others do. Every plant and animal has its own energy. For this reason alone, we ought not to damage them."
ERNESTO MOLINA, *Punta Chueca, instructing students*

We then realized that some of the children had very few chances for direct contact with some of the native species mentioned in Comcáac songs, stories, riddles, and tongue twisters. We began to take them out with knowledgeable elders on field trips along the Sonoran coast and even to Tiburón and San Esteban islands. There, in the very places where certain historic events gave rise to widely told stories about mythic animals and plants, elders would recount the episodes and sing songs to the very species celebrated in their oral traditions.

Figure 46. Educational materials developed by the Seri Ethnozoology Education Project, based at the Arizona-Sonora Desert Museum. At left is a booklet on traditional reptile lore distributed to Comcáac schools in 1998. In the middle are tapes of Seri songs about wildlife. At right is a booklet about native foods useful in preventing diabetes. (Photo David Burckhalter, 1999)

It was the first time many of these youths had an opportunity to visit the places associated with boojums, chuckwallas, and giant sea serpents.

"As the children are once again involved in direct contact with the native fauna, those who have the knowledge about these creatures will share the stories with the rest. It is important for all our people to know."

IGNACIO "NACHO" BARNET, *tribal governor*

Since then, we have worked with several Seri teachers on specific projects that take a variety of approaches to environmental education. The Columbus Zoo donated "turtle and tortoise discovery kits" to a new native-language immersion school begun in Punta Chueca. Each kit includes puppets, models, posters, and activities about native species and the threats they face. Then, with the help of ethnomusicologists Jack Loeffler and Thomas Vennum, Laurie Monti produced archival-quality recordings of Comcáac songs about animals as compact discs and cassette tapes, with a tape given to every family in Punta Chueca and Desemboque. Over a dozen Seri elders consented to being recorded so that their style of singing could be heard by younger generations. Historic tapings of deceased singers recorded over the last fifty years were also "repatriated," from the Smithsonian and Indiana University, to the Comcáac community for the first time. Another version of the recordings, entitled "Conservation Wisdom Hidden in Seri Songs," was distributed to all conservation organizations and agencies in the region; it includes a narrative by Laurie Monti explaining why the Comcáac should be involved in decision-making that affects the region's natural resources.

Under the direction of Laurie Monti, schoolchildren from Punta Chueca created a half-kilometer-long ethnobiology trail, featuring trailside signs for plants that their grandmothers identified as being important to Comcáac culture. Learning that Seri individuals have become increasingly susceptible to adult-onset diabetes, Monti (1998) also wrote a bilingual primer on how the consumption of traditional foods—from eelgrass and mesquite to turtle meat—can prevent diabetes by favoring the slow digestion of carbohydrates (see Fig. 46). In classes for Comcáac diabetics, she has used the Green Sea Turtle as an example of an animal that stays healthy by digesting its food slowly, thereby sustaining its energy over the long haul. Her primer on diabetes prevention was distributed to every Comcáac family in the autumn of 1998.

"*Moosni* [sea turtles] are like the Comcáac in that they are threatened by many things, [but they] have always survived, even when traveling long distances across dangerous waters. Both the *moosni* and the Comcáac know that the resources of nature [can] make them strong."

Translated excerpt from Los Comcáac y su comida tradicional: Como prevenir el diabétis, *edited by Laurie Monti (1998)*

Perhaps the most complex collaborative effort we have attempted is a captive breeding project based next to the Punta Chueca schools. It was originally suggested by Seri leaders and government officials when they visited the Arizona-

Sonora Desert Museum and were surprised to see how popular the chuckwalla exhibit was and how successful the captive breeding effort had become. They proposed that Seri individuals be trained not only in captive animal care but also in the monitoring and management of San Esteban Chuckwallas in their natural habitat. If they could attract enough ecotourism revenue to cover more boat trips out to San Esteban, they could then assume increased authority over protecting the chuckwallas and other island resources. At the same time, the schoolchildren would benefit from daily contact with the captive chuckwallas.

One month after a February 1998 training session for four Seri men at the Desert Museum, members of the Punta Chueca community helped us construct a brick-walled enclosure that surrounded an artificial mound of boulders, caves, and hibernation chambers. While both Seri and non-Seri men skilled in masonry and stucco finishing accomplished much of the technical work, it was a joy to see twenty local boys and girls hand-plastering the walls with the first coat of stucco and helping with the painting of signs. As the children painted animal images on signs, the elders debated the wording to be added, wanting to be sure that their message made sense in Spanish and English as well as in their native tongue. After much debate, the elderly men decided against calling the captive breeding enclosure a hiding place or hibernation site; instead it would be referred to as a place where chuckwallas are born and raised (Fig. 47).

Every step of the project was thoroughly discussed with community elders—where the animals should come from, who should go and get them, who should care for them, and what they should be fed. Botanical details of the project generated considerable discussion. When we transplanted forage plants into the enclosure in May—five months before the animals were introduced, to allow the plants to become established—we chose native species that we had seen on Isla San Esteban the year before. However, once the animals were in place, community members remembered other plants that they had seen chuckwallas eat in the wild, and urged us to add these species to round out the animals' diet. Several men insisted, for example, that chuckwallas love to eat the flower buds of cholla cacti, a food that herpetologist Charles Silber (1985) has also observed to be a favorite of island chuckwallas.

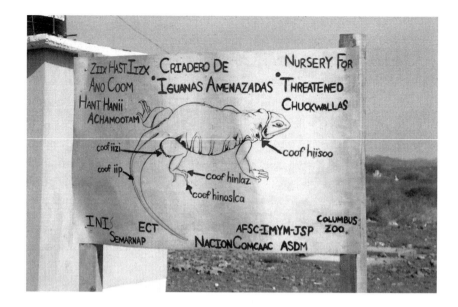

When one male and three female chuckwallas taken from Isla San Esteban in a Seri panga finally arrived in Punta Chueca, over forty children and adults joined in celebrating their homecoming in the community. Following a brief speech in Cmique Iitom by Alfredo López, four men and women released the animals into their new home amid cheers and squeals from the children (Fig. 48). Within minutes, each animal had a nickname, one of which facetiously likened its body shape to that of a community member. Each sunny weekday afternoon for the next few months, Seri parents would congregate at the chuckwalla enclosure and talk while waiting for their children to be released from school. The chuckwalla exhibit quickly became a gathering ground not only for observation and gossip, but also for the renewed transmission of ancient lore about animals. The founding chuckwallas have since reproduced in the exhibit, and their progeny may be offered to Mexican zoos as a means of furthering conservation education and providing income to the Comcáac who maintain the exhibit.

While the long-term maintenance of the animals on exhibit cannot be assured, the project has become one more step in a longer process: to ensure that

Figure 48. Alfredo López releasing chuckwallas into a captive breeding enclosure area next to the Punta Chueca school, September 1998. (Photo Gary Nabhan, 1998)

the Comcáac community remains involved in the utilization, protection, and celebration of its reptilian neighbors. Each phase of the project has attempted to reinforce the value of Comcáac knowledge regarding the natural world, rather than assuming that academically trained scientists can provide all the pertinent information the tribal and federal governments will ever need to guide the management of the region's wildlife. To reiterate the words of Italian folklorist and novelist Italo Calvino (1983), "New knowledge . . . does not compensate for knowledge spread only by direct oral transmission, which, once lost, can never be regained or retransmitted."

This realization has led to an initiative to blend traditional ecological knowledge and conventional methods in field biology to better inventory, monitor, and protect threatened species in Comcáac territory. For many months, Punta Chueca resident Cornelio Robles expressed concern that most field surveys for wildlife conservation on the islands paid Seri participants only as boatmen or trail guides, even though they had begun to assist biologists with making observations, measurements, and assessments of various species' status. During a

December 1998 visit to the Arizona-Sonora Desert Museum, Robles requested the assistance of the museum and other conservation organizations in certifying Seri individuals who not only retained their community's traditional ecological knowledge but had also become competent in field techniques used by conservation biologists.

We then developed a "*para-ecólogo*" certification course, comprising six modules involving identification, monitoring, and protection of different sets of threatened plant and animal species, in which Seri individuals are trained by both traditional elders and Western scientists. With this training, which is endorsed by several academic and government institutions, the Comcáac community has begun to receive direct support for endangered species monitoring and management, rather than seeing all such support being funneled to universities and research institutes. At the course's inauguration, elder Genaro Herrera told the trainees that by learning how to defend the riches of Seri territory they would become the equivalent of "warriors" of past ages; their weapons would not be bows or guns, however, but empirical knowledge and skills in natural resource management (Fig. 49). He then sang an iquimuni, "war song," to the encircled youths, thus initiating them into the status of warrior-trainee. Twelve para-ecologists have now graduated and are involved in managing their own wildlife monitoring and recovery projects.

If more Seri young people can obtain work in the natural resource management field, this may provide one more way that traditional ecological knowledge will remain relevant to those making a living along the Sonoran coast. In addition, it will allow young people to come regularly into contact with the lizards, turtles, and other reptiles that have long been part of their people's lore, their ceremonies, and their well-being (Nabhan et al. 1999). During the fall of 2000, a traditional school was inaugurated in Punta Chueca by elder Antonio Robles as one more means of providing young Seri individuals with training in the native flora and fauna and their curative powers. The opening of the school was celebrated with a feast of traditional foods, including fish, reptile meat, prickly pear, mesquite, and agave.

Perhaps the children attending this school will be more prone to heed the

Figure 49. Tribal elder Genaro Herrera blessing a bird caught and then released by para-ecologists. Most Seri wildlife-monitoring efforts involve capture, marking, weighing, and release, with the hope of future recaptures. (Photo Kit Schweitzer, 1999)

words of their former tribal governor, Pedro Romero, when he said that "Islas Tiburón and San Esteban and the midriff of the Sea of Cortés have great economic and cultural significance for us, the Comcáac. It is from here that our people obtained an important part of our nutritional and spiritual sustenance, and received the strength to survive times of great social or natural adversity."

I believe that Comcáac youth will also receive such strength from contact with the native reptiles of their homeland. As I departed from Punta Chueca after one extended stay there, my memory carried an image that embodies such hope: that of my one-year-old namesake, Hant Coáaxoj, sitting in the sand in front of his grandfather's house facing Isla Tiburón, playing with a small effigy of a Regal Horned Lizard.

part two

accounts of reptiles known by the comcáac

IN PART 1 of this book, I have used various devices—personal narrative, conversation, history, and theoretical discussion—to present the ecological knowledge of the Comcáac people and how it relates to current trends in Western science. Part 2 (chapter 9) is perhaps a more "standard" ethnobiological monograph in that it offers a species-by-species account of the herpetofauna of the central Sonoran coast and its adjacent islands. The information presented here—a status report, in essence—combines traditional and practical Seri lore and knowledge, derived from generations of cultural reliance on these animals, with Western scientific systematics. Neither of these frameworks is assumed to be more or less "correct" than the other; they are simply different takes on an array of distinctive animals. Together the Seri ethno-systematics and the Western classifications provide a fuller understanding of species that have been, and remain, of central importance to the Comcáac people.

TURTLES AND TORTOISES

TESTUDINIDAE: Land Tortoises

Gopherus agassizii
English common name: Desert Tortoise
Spanish name: *tortuga del monte*
Comcáac names: *xtamóosni, ziix hehet cöquiij, ziix catotim*

The archaic name *xtamóosni* for the Desert Tortoise is curious linguistically. Hine (2000) was told that this name may mean "earth turtle," possibly derived from *xta*, which he was told was a contraction of *hant*, "earth/land/place/present time," and *moosni*, "Green Sea Turtle/sea turtle/generic turtle." If this is not correct (which the Mosers [in prep.] suspect), it is perplexing why a desert dweller would be associated through the prefix *xta* with mud and lagoons, as noted below in the mud turtle discussion.

The Comcáac hardly ever use this term today, except when singing traditional songs or referring to forage plants associated with Desert Tortoises called *xtamoosnóohit,* "Desert Tortoise forage" (Felger, Moser, and Moser 1983). Instead, most Comcáac interchangeably use the two descriptive nicknames for the Desert Tortoise, *ziix hehet cöquiij,* "thing sitting under plants," and *ziix catotim,* "thing that slowly scoots along."

Most Comcáac are familiar with the times of Desert Tortoise hibernation and estivation both, which they call *cpoin,* "the closing-in." They say that the tortoises enter shallow burrows under trees and in arroyo banks, caves, and, sometimes, unsheltered depressions known as pallets. They retire to these pallets and burrows in November, and emerge from them as early as March and as late as May; they also use them during the hottest months. Comcáac elders report that the caves and shallow burrows in which Desert Tortoises hibernate will also harbor commensals such as Gila Monsters, Desert Spiny Lizards, Coachwhips, rattlesnakes, and packrats (Nabhan 2000b). Bury et al. (in press) likewise note that the shelters dug by Desert Tortoises become cover for many other reptiles, invertebrates, packrats (*Neotoma albigula*), and even for burrowing owls (*Athene cunicularia*). Guadalupe López led us to a shallow burrow at Ensenada del Perro which contained not only a Desert Tortoise but also a Western Diamondback Rattlesnake.

One early report states that the Comcáac believe Desert Tortoises are gregarious and stay in family groups (Malkin 1962). Comcáac elders have reported to me that two to three tortoises may live together, especially pairs of a male and a female. They have also seen males fight one another, particularly when "one of their ladies is nearby." Some believe that while each tortoise has its own territory, it may cover considerable ground during its diurnal meanderings. Early reports that all Desert Tortoises in Sonora were nomadic with large activity ranges have been tempered by studies from Isla Tiburón that demonstrate allegiance to a number of burrows and pallets dug in a relatively small area (Reyes-Osorio and Bury 1982; Bury et al. in press). These studies, aided by the local knowledge of Desert Tortoise burrows by Seri field technicians Alfonso Méndez, Oswaldo Méndez, Guadalupe López, and Chapo Barnet resulted in density estimates of twenty-nine to eighty-seven Desert Tortoises per square kilometer, which are comparable to the highest densities of tortoises recorded in the more favorable parts of the Mojave Desert. Directed by Mercedes Vaughn, surveys by Comcáac para-ecologists and Desert Tortoise Council researchers have reported densities of ten to twenty live animals per square kilometer, with high numbers of tortoises killed by drought in the late 1990s on both the mainland and Tiburón (Vaughn 2002).

The Comcáac claim that they can hear Desert Tortoises vocalizing during mating from some distance, that the cries of a male while mounting a female in one canyon can be heard from the next canyon over in the same mountain range. They have consistently suggested that females lay eight to ten eggs per mating, from the 1950s when first questioned by Malkin (1962) to the present.

At least six plants are known to the Comcáac as tortoise forage, including *Chaenactis carphoclinia, Chorizanthe brevicornu, Allionia incarnata, Fagonia californica,* and *F. pachyacantha* (Felger and Moser 1985; Nabhan 2000b). Recently, I was told by an elderly Punta Chueca resident that another plant, known to the Comcáac as *xomítom hant cocpétij* and to researchers as *Trianthema portulacastrum,* was an important tortoise forage in the area. Later, other Seri men volunteered the same information when encountering this false purslane near Tecomate. This information is important if only because none of the twenty-some published studies on the diet and nutrition of Desert Tortoises have been carried out in northern Mexico — even though 33–40 percent of the species' range occurs in Mexico — and only 2 percent of the walking surveys of habitat use by tortoises have been done south of the U.S. border (Bury et al. in press).

The Comcáac have observed several native animals preying on Desert Tortoises in their homelands, including coyotes, bobcats, mountain lions, and hawks. Whereas Malkin (1962) recorded that the Comcáac believed that mountain lions were the only effective predator on tortoises, at the scene of the death of one Desert Tortoise, Ernesto Molina inferred that a hawk had struck at it. Another observer commented that coyotes are sometimes ineffective at predation because the tortoises can retreat into their shells, which coyotes are unable to penetrate either with their bite or with their claws.

The Comcáac know that birds, snakes, and lizards, as well as humans, prey upon tortoise eggs. These eggs were opportunistically gathered and eaten by the Comcáac in the recent past. Although contemporary Comcáac adults mention that tortoise eggs have been a traditional food during their lifetimes, the practice is declining (Nabhan 2000b).

Most Comcáac over fifty years of age claim to have hunted and eaten Desert Tortoise at one time or another, but it is my impression that the timing and location of tortoise capture and consumption were culturally regulated by a variety of taboos. Despite hundreds, and perhaps thousands, of years of Comcáac utilization of tortoises on Isla Tiburón up until its abandonment in the 1940s, demographic surveys there suggest that today the island has one of the densest Desert Tortoise populations known from the Sonoran Desert (Reyes-Osorio and Bury 1982). As Felger, Moser, and Moser (1983, 117) conclude, "Seri hunting pressure, even in earlier times, does not seem to have been a major factor affecting tortoise populations, since there was at least as great Seri population density on the island as on the mainland."

The Comcáac as well as the neighboring O'odham believe that the Desert Tortoise has the power to make disrespectful humans sick (Bahr et al. 1979; Nabhan 2002). Even though men and women may capture Desert Tortoises as pets or for food, they cannot do this at all times and in all places. For instance, a Seri out hunting Mule Deer on Isla Tiburón is not allowed to capture or kill a Desert Tortoise during the hunt, for it will bring bad luck.

Pregnant women are discouraged from hunting tortoises. If a woman gives birth to only female offspring, the reason given is that she must have eaten the reproductive organs of a female tortoise; if her children are all male, it is because in her youth she was hit in the small of the back with the penis of a male tortoise thrown at her by one of her girlfriends (Felger, Moser, and Moser 1983).

However strange such beliefs sound to outsiders, they will likely curtail a person's willingness to hunt tortoises. It may be worthwhile to consider this as one of many reasons Isla Tiburón's tortoises are so abundant.

Comcáac girls historically made clothing for their clay-figurine dolls with the dried, softened fragments of tortoise bladders (Felger, Moser, and Moser 1983). A medicinal application of Desert Tortoises exists as well: poor eyesight can be cured by heating a stone in the body cavity of a tortoise, putting one's face over the resulting steam, and opening one's eyes to the steam.

There are many Comcáac songs about the Desert Tortoise. In one, a tortoise sees so many wildflowers up in the hills that it laughs at the notion of so much food being available in the desert. Two other tortoise songs are loosely translated below (they are discussed in greater detail in Nabhan 2002).

Los rayos del sol	The rays of the sun
están pasando	are passing
atrás de la tortuga.	behind the tortoise.
Siempre anda	Always walking,
la tortuga del desierto	the Desert Tortoise,
en una dirección	in a direction
fuera de la puesta del sol.	away from the setting sun.
Una tortuga estaba caminando en un cerro	A tortoise was walking on a hill
cuando se faltó en el cerro	when he broke down
tratar de comer algo.	where he was trying to eat something.
En aquel entonces, los cazadores lejanos	Then hunters from afar
llegaron para matarlo.	came to kill him.
Ya hay peligro—	Now there is danger—
tiene que esconderse.	he has to hide.
Hay peligro si los cazadores	There will be danger if the hunters
están viendolo—	get to see him—
tiene que esconderse.	he has to hide.

KINOSTERNIDAE: Mud and Musk Turtles

Kinosternon flavescens and K. sonoriense
English common names: Yellow Mud Turtle and Sonoran Mud Turtle, respectively
Spanish name: tortuga de los charcos
Comcáac name: xtamáaija

The name xtamáaija includes the prefix xta-, which may signify an association with mud, moist earth, or estuaries and lagoons. It is also part of a compound lexeme, xtamáaija oohit, "mud turtle's eatings," which refers to the herb Nemacladus glanduliferus, a forage that mud turtles eat in moist, grassy places. Jesús Rojo has found mud turtles in seasonal waterholes along intermittently flowing watercourses, in irrigated areas where tail waters accumulate, and in canals where water regularly flows during the warm season. Antonio López associated these turtles with puddles of "sweet [fresh] water" that appeared after the summer rains.

All Seri individuals interviewed who knew any mud turtle referred to it by this name alone and did not recognize the subtle differences between the two Kinosternon species, with distributions extending into their territory, though seldom on the coast. These species are seen together at the same waterhole at only a few localities north of the contemporary range of Comcáac travelers, such as the Quitobaquito and the Quitovac oases in O'odham territory. Tom Van Devender (pers. comm.) has found a dead Yellow Mud Turtle in Arroyo San Ignacio north of Desemboque, although this should be the rarer species in the central Gulf coast region of Sonora.

Only a few of the Seri individuals who had lived inland acknowledged that mud turtles were once eaten. Mud turtle consumption by any of the Comcáac is somewhat surprising, since other indigenous peoples avoid these animals because of their smell. Mud turtles use their "stink pots" as a defensive mechanism to discourage predators.

In a beautiful song chanted to entice summer rains to the Comcáac homeland, a giant xtamáaija is called upon to bring fresh water with him to the parched land. As Adolfo Burgos explains the lyrics, a hunter wandering out in a dry valley was close to dying of thirst. He called upon a person with mud turtle spirit power, promising to pay him with three deer hides if the rains came. As *xtamáaija*'s spirit power became activated, a cloudburst brought on such intense rains that it looked as though the entire world was being flooded. The hunter happily drank this fresh water, his life saved.

CHELONIIDAE: Sea Turtles

Caretta caretta
English common name: Loggerhead
Spanish names: *caguama cabezona, jabalina*
Comcáac names: *xpeyo, moosni ilítcoj caacöl*

While the word *xpeyo* is unanalyzable, *moosni ilítcoj caacöl* means "turtles big-headed" (as does the Spanish term, *caguama cabezona*). These large-headed sea turtles, which are known in two forms, are not considered particularly flavorful or otherwise desirable as prey items. The first, more common, form, *xpeyo*, has a reddish brown carapace that tapers toward the back and may achieve weights of 80 to 150 kilograms. While some say that Loggerheads have been rarely seen in the Canal del Infiernillo since the 1940s, Jesús Rojo claims that at least five juveniles were observed north of Bahía Kino since 1995 and that they remain at least as common as Hawksbills along the Sonoran coast.

The second form, *moosni ilítcoj caacöl*, is typically spoken of in the plural. Ramón Perales says they are not very tasty compared to *xpeyo* and have greener, smaller heads as well as "milder manners." Others say that these turtles have yellowish-cream–colored plastrons and are quite massive. They were historically seen from Bahía Kino southward, but since the mid-1950s the Comcáac have encountered them only infrequently.

Ramón Perales says that the Loggerhead behavior is not that different from the behavior of various Green Sea Turtle races, except for their general lack of fierceness. They are easy to bring aboard boats compared to other turtles of similar size. José Juan Moreno sang a sea song about *xpeyo* that Ernesto Molina roughly translated as follows: "When a Loggerhead needs a breath below, it rises for air to the sea's very surface, doing so just as a strong wind gusts up, forcing it to dive to the depths of the sea."

Chelonia mydas
English common name: Green (or Black) Sea Turtle
Spanish name: *caguama carrinegra*
Comcáac names: *moosni, ziix xepe ano quiih*

The term *moosni*, possibly an ancient Seri term for land and water turtles, was perhaps borrowed by the Uto-Aztecan neighbors of the Comcáac, the Mexican Yoemem and O'odham, to refer to sea turtles once these people began to reside on the Sonoran coast. The archaic descriptive phrase *ziix xepe ano* quiih, "thing that is in the sea," may have been a nickname given out of affection and familiarity, much as was done with toads and Desert Tortoises.

The Comcáac use their native term *moosni* as well as the Sonoran Spanish term *caguama* both distributively for all sea turtles and specifically for the Green Sea Turtle. Because of the polysemy of these terms, it is often difficult to determine whether statements by early chroniclers are generic or specific. While McGee (1898, 214), for example, clearly specified that at the Seri camps he visited "the most conspicuous single article is the local green sea turtle," he typically spoke of sea turtles in a generic way.

McGee (1898, 290–291) likened sea turtles' importance to the Seri to the importance of bison among the Sioux, claiming that they carried material culture fashioned from turtle bodies with them from birth to death. He noted in particular that "turtle bone dolls were put in burials along with personal possessions; on top of pelican pelts, two turtle shells [were] laid over as a kind of coffin. Mortuary food includes portions of turtle-flippers and a chunk of plastron—food for the journey."

On the basis of his limited observations, McGee (1898, 213) guessed that sea turtles made up one-quarter of the Seri diet, an estimate discussed by Malkin (1962), Felger and Moser (1985), and other later scholars without radical revision. Most observers agree that sea turtles were historically a defining element of the diet and material culture of coastal Comcáac and that the depth of their material and spiritual dependence on these animals can hardly be overstated (Smith 1974; Cliffton, Cornejo, and Felger 1982; Felger and Moser 1987). Sea turtles were hunted throughout the year, though most species and populations were more abundant during the warm season (Cliffton, Cornejo, and Felger 1982). Comcáac hunters had ample opportunity to observe turtles mating, migrating, hibernating, and feeding. The hunters' primary tool was this detailed knowledge of sea turtle behavior and habitat preferences, but over the last two centuries they also incorporated numerous technologies—from ironwood and iron harpoon points, nylon filament nets, and hookahs to motorized pangas—in their pursuit of sea turtles.

The demand for sea turtle meat in Mexico offered good wages to many Seri between the 1940s and the 1970s, but this commodification did not break the nutritional and spiritual links between the Comcáac and Green Sea Turtles. The sale of sea turtle meat to outsiders gradually engaged more and more Seri families in the 1930s and 1940s and by the late 1960s had outstripped other sources of cash income to their communities. Intense competition from other turtle hunters, and incidental take by long-lines, shrimp trawlers, and chinchorro nets forced the Comcáac to all but abandon sea turtle hunting as a source of income and regular sustenance by the late 1980s.

In the late 1970s, Arizona-Sonora Desert Museum staff began to lobby on behalf of the Comcáac community (and on behalf of turtle species) to stop the slaughter of sea turtles elsewhere in Mexico, hoping that at least a few turtles would continue to arrive in the Sea of Cortés each year (Cliffton, Cornejo, and Felger 1982; Cliffton 1990). Around the same time, overexploitation of all sea turtles in the Pacific compelled Mexican government officials to limit commercial harvesting of every turtle species that entered its waters. Seri ceremonial use of Green Sea Turtles was later exempted from this ban but is apparently limited to four animals per ceremonial event (although a few Seri claim that no more than four animals are captured per year for puberty ceremonies).

Whether the meat was sought by the Comcáac for local consumption or for sale to outsiders, it was hunted and handled in much the same way. A harpooner stationed near the prow would search for surfacing sea turtles, sometimes maintaining this position for six to eight hours or for thirty miles of hunting along the coast of Isla Tiburón. Upon glimpsing one, he would direct the pilot to move the boat into the bubble trail left by the turtle's last expiration. When the turtle surfaced next to expel air, the harpooner would ready himself. As the animal dipped back down, it would be close enough to the surface that the harpooner could see its form, and when it resurfaced, the harpooner would drive an ironwood or metal point into its carapace or flipper.

The Comcáac developed different strategies for hunting turtles by full moon, by lamplight, by bioluminescence, and by searching rocky reefs by day in the summer or sandy dormancy grounds by day in the winter. During the summer, men often hunted turtles at night, using lights fueled by white gas to attract them. Another hunting strategy was employed during new moons when bioluminescent organisms floating at the sea surface made it obvious where turtles broke through the "phosphorescence." Guadalupe López could spot a turtle's trail through bioluminescent tides at a distance four to five times greater than that by which he could locate them by daylight, leading to his fame for being able to "see in the dark."

Once the turtle was harpooned, but before it was lifted into the boat, a slit was sometimes made in its neck and its fat sampled, as observed by ethno-biologist Alfred Whiting (1951). The turtle was then brought onto the deck by the harpooner and the "gaffer," or helper, who grabbed its front flippers while the pilot or motorist steered the boat. Although many turtles were transported alive, they were sometimes clubbed on the nose and killed on the spot. In either case, they were flipped on their backs until the boat was beached.

Back on land, the women were usually the ones who cleaned and butchered them, working with the animals placed belly up in the intertidal zone. The turtle's throat was cut with a knife, and its crop, stomach, and intestines were then extracted. Sometimes the entrails were offered to the camp's dogs. Turtle bladders and stomach linings were cleaned, tied off at one or both ends, and used as vessels for water, oil, and other materials (Villapando 1989; Felger and Moser 1985), while intestinal fibers were used as fiddle strings for the Comcáac monochord violin.

Once killed and eviscerated, the turtle was prepared for eating by building a fire on its plastron, usually fueled by branches of *seepol* (*Frankenia palmeri*). After roasting, the plastron was removed and the meat immediately beneath it offered as a "first feast" to those lucky enough to be present. The rest of the meat, together with the blood, oil, and other juices, was then removed, with chunks of meat temporarily placed atop mounds of *moosni iti hatéepx* (*Croton californicus, Abronia villosa,* and *A. maritima*) to keep it off the sand (Felger and Moser 1985). These foods and drinks were then taken to nearby camps and ritually shared with others. Meat and juices from turtle flippers were the only foods prepared for men to ready them for paddling balsa boats to San Esteban from Tiburón or elsewhere. William N. Smith (1951) reported that boys were not allowed to eat any of the meat from the first sea turtle they killed, although a fiesta was held on their behalf; they could resume eating sea turtle meat eight days later. Smith also saw sea turtle blood used in the 1940s as face paint by Seri men and boys, who drew horizontal or vertical lines across one another's noses or butterfly wings on one another's cheeks.

While Felger and Moser (1985) do not list sea turtle eggs as part of the Comcáac diet, one elderly Seri woman reported to me that she had once eaten the eggs from a gravid female that was killed inadvertently. Becky Moser (pers. comm.) told me that this practice was routine through the 1950s and 1960s; her daughter Cathy, she said, once joined several Seri women in rescuing eggs to eat after a sea turtle being butchered was found to be a gravid female. There was no point in letting the eggs go to waste. Although a few contemporary Seri have witnessed Pacific (Olive) Ridleys laying eggs on sandy beaches, they did not eat any of these eggs.

Once the butchering was completed, the turtle carapace might be left to dry, to be reclaimed later for use as a windbreak, as thatch, as a cradle, or as a container. Hawksbill carapaces are sometimes elaborated into crafts for personal adornment or for sale. In 1998, a Hawksbill accidentally killed in fishermen's nets was retrieved, and its shell was made into a quail figurine by an elderly Desemboque woman.

Traditionally, sea turtle bones might be dressed up as dolls for baby girls, but bone effigies have also played an important role in Comcáac ritual observances. Sea turtle clavicles, humeri, and femurs were disarticulated, scrubbed, and placed in the sand as effigies in ceremonial groupings. Some of these bone figurines have the same shape as anthropomorphic (female) figurines of clay or of ironwood still elaborated by contemporary Seri artisans. Anthropologist William N. Smith (1951, 1974) associated this ceremonial use of sea turtle bones with the creation stories relating a giant sea turtle's role in creating land for the Comcáac.

Although their ceremonial life was not as calendrically structured as that of many of their agricultural neighbors, the Comcáac did associate certain moons with the arrival and with the peak abundance of Green Sea Turtles. *Caayaj zaac*, their name for the second moon after winter stolstice (around February), for example, means "moon of few [*cooyam*] hunters." It marks the first month following the arrival of young Green Sea Turtle migrants (*cooyam*) coming into the Sea of Cortés for the first time in their lives. *Cayajáacoj*, "moon of many hunters," comes later, when the migrants were so abundant that Seri men could harpoon them from the shallows in front of their beach camps.

A person who gained "sea turtle spirit-power" through a vision quest or by some other means was called *moosni quiho* and was greatly esteemed.

Felger and Moser's (1985, 42–49) lovely summary of Comcáac interactions with sea turtles remains the best introduction to this topic, covering both generalities and particulars. Many important details of turtle ethnozoology are not amply discussed in that brief summary although they can be found in other accounts as well (Cliffton, Cornejo, and Felger 1982; Felger and Moser 1987; Smith 1974). Elsewhere Felger and Moser (1974; 1985, 113) detail medicinal uses of Green Sea Turtles, such as the herbal tea made of dried sea turtle penises, drunk to induce conception, which they claim the Comcáac still used on occasion through the 1980s. They also report that pieces of dried turtle intestine were boiled in a small amount of water and drunk as a remedy for seasickness and, more recently, for car sickness. While most sea turtle populations or races might be used for this purpose, other remedies require a particular race; only the oil of young migratory *cooyam* was considered effective as a treatment for craziness. Turtle intestinal sinew, carapaces, and bones also have various uses (Felger and Moser 1985; Felger 1990).

There are many terms describing sea turtle morphology and anatomy (Moser, Moser, and Marlett in prep.). A preliminary graphing of morphological terms for sea turtles is presented in Figure 16; it includes commonly used terms for carapace (*ipócj*), for plastron (*iti ihímoz*), for the gelatinous cover on the flippers, carapace, and plastron (*iixt*), and for additional morphological features. Other traits, such as a strange skunky fluid called *isa*, released from the gray, lumpy Rathke's glands of *cooyam* migrants and possibly functioning as a shark repellent, have only recently been investigated scientifically (Felger and Moser 1987).

The Comcáac still use at least fourteen terms to refer to what others consider to be races, populations, developmental stages, and/or legendary individuals of Green Sea Turtles (Felger and Moser 1985, 1987). Felger and Moser (1976) subsume these fourteen kinds of sea turtles under the single species of *Chelonia mydas* recognized by scientists in the Pacific Ocean, claiming that each constitutes a distinctive "microrace," or subspecific gene pool of sea turtles. From my own interviews with elderly Seri sea turtle hunters I conclude that certain folk taxa are the equivalents of cohesive populations or species within the broader taxon of *moosni*, whereas other terms apply to habitat-specific developmental stages, recessive or hybrid forms traveling within more common forms, or mythic animals, as either individuals or species. Kim Cliffton (pers. comm.) has seen all but the mythic blue turtle, confirming that these thirteen other entities still occur in the Pacific.

In the following treatment, I have asterisked those terms which Comcáac consultants clearly use for full-fledged races of Green Sea Turtles; the remaining terms describe levels of genetic, developmental, behavioral, or phenotypic variation. In one case, which sea turtle physiologist Scott Eckert helped me interpret, the Comcáac recognize one migratory race of Green Sea Turtles by three names associated with different phases in its physiological development, each having to do with the capacity to dive and forage in waters of particular depths and temperatures. Known simply as *cooyam*, recently hatched migrants entering the Canal del Infiernillo for the first time stay in warm, shallow waters so close to the coast that people could walk out among them. Later, as pre-reproductive juveniles migrating in for a second or third time, they are seen in a wider range of habitats in the Canal and are called *cooyam caacoj*. Then, as mature, reproductive individuals, the fattened *ipxom haaco iima* have enough insulation to dive to greater depths in the trough between Islas San Esteban and Tiburón or off the northern and western coasts of Tiburón; they now add jellyfish to their youthful diet of algae and eelgrass. As Guadalupe López once warned a few younger Seri men in my presence, it is hardly worth lumping all these named variants under one term, since they will be found in different areas, behave differently, and even taste different, depending on their diets.

Future ethnobiological investigations among the Comcáac should attempt to identify whether the named color variants are rare individuals traveling among otherwise uniformly colored groups or whether they are found in isolation, under specific conditions. Such research will hopefully also elucidate which sea turtles the Comcáac are referring to when they speak of *stac casíi*, a kind of turtle with a peculiar smell (perhaps due to its seasonal diet). It is clear that the Comcáac are naming phenotypes that biologists in the Sea of Cortés have hardly studied, even though researchers in the Caribbean and in Australian coastal waters have confirmed that identification of these phenotypes provides a valid means of understanding intraspecific variation and habitat-dependent developmental sequences.

There are many songs about all kinds of *moosni* within the musical genre called *xepe an cöicóos*, particularly about large male turtles seen by moonlight or in phosphorescent seas by harpooners. One song recounts how harpooners would thump the bottom of their boats with harpoon shafts to scare the turtles into coming up and breaking through the glow. In May 1999, when the Comcáac Council of Elders took sea turtle conservation biologists and para-ecologist trainees out into the Canal del Infiernillo to search for sea turtles, no turtles appeared until José Juan Moreno, the eldest, sang one of the *xepe an cöicóos*. The first Green Sea Turtle was spotted within a minute after he began singing.

Moosni áa, "real (common) sea turtle." This term serves to distinguish Green Sea Turtles from the other species of sea turtles, which collectively are referred to by the generic, distributive term *moosni*. Below are the named variants within what scientists believe to be the *Chelonia* gene pool.

*Moosnáapa, "true (authentic) sea turtle." This term refers to the mythic sea turtle originally eaten by the ancestors of the contemporary Comcáac; it is the one that helped create the land on which the Comcáac live today. It is generally said to have brownish gray flippers and carapace, the latter having clearly incised plates reminiscent of those of Hawksbill Turtles, except not raised and separately articulated. Guadalupe López says that its coloration ranges from grayish brown to grayish green, and that it has one claw per flipper.

López and Ramón Perales say that this race still exists, whereas Jesús Rojo and others lament that it is no longer seen. Ramón contends that one sea turtle brought into Desemboque in March 1998 fit this category, for it had a carapace tightly fitted to the body of the animal and its plates were well outlined. He also claims that it is the best tasting of all sea turtles, as well as being the first kind ever eaten by the Comcáac. Adolfo Burgos sings one *xepe an cöicóos*, "ocean song," about it swimming at night through the phosphorescence.

Moosníil, "bluish sea turtle." This mythic variant no longer exists. According to Ramón Perales, it was seen only by his ancestors. Jesús Rojo says that it migrated away, never to return, while others claim that it was formerly seen only off the northern and western shores of Isla Tiburón. Both huge and strong, it earned its name from its magical capacity to turn both harpoon points and mesquite ropes blue when it was speared and captured. This sea turtle's blueness is attributed to its eating of the feathery red alga *Asparagopsis taxiformis*, called by the Seri *moosníil ihaquéepe*, "blue turtle's favorite [delicacy]." When María Antonia Colosio was shown this alga by Felger and Moser (1985) she said, "The blue turtle eats it—you know, the one the blue fluid comes from."

Guadalupe López claimed that its coloration was lead black, while others claimed it had a bluish cast as well. In addition, López remembered that it weighed in the 35–45 kilogram range and had one claw per flipper; gravid females, he said, carried soft eggs. It is the sea turtle referred to in a harpooning song widely known among the contemporary Seri, one that Jack Loeffler recorded being sung by Jesús Rojo:

cotíin, cotíin eah,	I am thrusting my harpoon down, hear its sound?
cotíin, cotíin eah,	I am thrusting my harpoon down, hear its sound?
cotíin, cotíin eah,	I am thrusting my harpoon down, hear its sound?
cotíin cotíin eah,	I am thrusting my harpoon down, hear its sound?
moosníiali, moosníil-cojíit,	Turtle sea-blue, turtle dropping down,
cotíin, cotíin eah,	I am thrusting my harpoon down, hear its sound?
cotíin, cotíin eahhh	I am thusting my harpoon down, hear its sound?

Moosníctoj, "reddish sea turtle." This rare albino, a pale pink or orangish variant, has been seen at the south end of the Canal del Infiernillo; nearly all the contemporary hunters at Punta Chueca know of its resting ground there, *moosníctoj iime.* Some say that its coloration is close to that of the *xpeyo,* or Loggerhead, but that its head is not as large. It has one claw per flipper. It was eaten at a ceremony in Punta Chueca in 1998.

Quiquíi, "weak" or "shriveled." A morphological variant with sunken eyes, a caved-in plastron, and thin, nearly motionless flippers, this sea turtle is known to have had little meat but considerable fat. It was occasionally seen in the Canal through the 1950s, but Jesús Rojo claims it no longer occurs in the area. This may have been a diseased sea turtle, affected by one of the several microbiotic blooms that occasionally kill marine mammals, reptiles, and fish in the Sea of Cortés (A.-L. Figueroa, pers. comm.).

Cooyam, "pilgrim" or "migrant." Used to describe the earliest and youngest of the migrants seen in the spring in the Canal del Infiernillo, this term implies that there is at least one population of Green Sea Turtles that does not overwinter in the Canal but is born to the south of Comcáac territory; it later returns as *cooyam caacoj*, then as mature **ipxom haaco iima*. When this migratory population enters the Canal for the first time in late February and March, forty to seventy small turtles are often seen swimming together close to shore. Some Seri turtlers say that they do this to reduce the chances of attack by predators. This population is known by its pale plastron and dark amber carapace (Seminoff et al. 2002).

The earliest arrivals, says Ramón Perales, are only 30–34 centimeters long. They typically do not have much meat on them yet, and their guts are nearly empty, but they still retain some of their baby fat. Ramón remembers that the first turtle he ever touched as a boy was a *cooyam* (he called it a *caguamita*) that had come in near the shore at Campo Viboras. There, his family could get up on the dunes or beach cliffs and watch them as they traveled in groups, coming up for air and diving back down in unison, like pods of dolphins. Ramón added that Pacific Ridleys, Hawksbills, and Leatherbacks never travel in groups.

If puberty ceremonies occur in late winter or early spring, it is likely that this is the sea turtle prepared for the young celebrants. *Cooyam* fat and oil also have curative uses, especially for youths and for those who have psychologically "strayed." The fat of *cooyam* is the only sea turtle fat that can be used to cure a crazy person. It is burnt where the distressed victim can sit in the smoke; as a Seri individual told Felger and Moser (1985, 113), "Make him stay in it, he will get well." During male initiations, if a youth suffered psychological trauma as part of his vision quest, the elderly initiators would rub his head with fried *cooyam* grease. Interestingly, the term *cooyam* can be used to refer to any individual, human or animal, on a long and arduous journey, whether it is a migratory turtle in the sea, a bighorn sheep moving across a mountain pass, or a

seeker of visions. Perhaps this ministration for a youth who feels like he is losing his mind is a way of providing the protection of another pilgrim, allowing
him to proceed on his psychologically risky vision quest.

Cooyam caacoj, "large pilgrim" or "big migrant." As they grow and gain weight, juvenile migrants are referred to by this name when seen migrating northward some
time after the initial arrival of hatchling *cooyam.* They forage in a wide range of
waters, but when they become sexually mature and capable of diving in the
coldest, deepest waters, they become known as *ipxom haaco iima.*

*Ipxom haaco iima, "one who is corpulent." As the mature phase of migrants, these
turtles are typically encountered near the northern, western, and southwestern
shores of Isla Tiburón, in deep cool waters, where they feed on *copsíij,* "sea gooseberries" (ctenophores), and jellyfish in addition to algae and eelgrass. They feed
on the jellyfish with their eyes closed, to reduce the stinging sensation resulting from these stinging animals' contact with sensitive skin and blood vessels.
Jesús Rojo claims that these fat-bellied sea turtles are still frequently seen around
the coasts of the midriff islands. Ramón Perales says that to be classified as this
morphological or development phase, sea turtles not only must be fat but also
must be at least ten to fifteen years old.

Moosni quimoja, "dying (?) sea turtle." Ramón Perales claims that this giant sea turtle was seen only once, just before it was to die; it is unclear to me whether
this event occurred during mythic or ordinary times.

Moosni ctam hax iima, "sea turtle male with hardly any testes." This morphological
variant is analogous to the developmentally arrested Mule Deer bucks on Isla
Tiburón called *hap imítjc,* with antlers that never pass the velvet stage and with
testes that never "drop." Many Comcáac seafarers have seen these very fat males
with short tails and undeveloped testes, but they have never seen them mating.
Jesús Rojo called one *medio macho,* "half a buck," because it lacked a scrotum.
Ramón Perales observed that these male turtles hardly grow tails, so at first glance

they almost look like females. Because they don't exert energy in sexual pursuits, they are never skinny, but are rich in fat. Several potential explanations exist for why some male turtles have arrested sexual development—diet-driven, stress-induced, or genetically predisposed hormonal imbalances, for example. The fact that the Comcáac have witnessed this condition frequently enough to name it indicates that is is worthy of more scientific inquiry.

Moosni hant coit, "sea turtle touching (bottom) ground." This population is peculiar to the Canal del Infiernillo and its shoals. These nonmigratory turtles establish iime (gathering grounds) that are used as winter-dormancy resting areas as well as summer retreats. Their wintering rests are three to five meters apart on shoals or in adjacent eelgrass beds; collectively, they form a moosni iti ihom, "mass of sea turtles." Some of these turtles may go dormant as early as November, but according to Guadalupe López most do so in December, regardless of how warm or cold the waters have been. The females first congregate at somewhat rocky sites, where they are inoculated with algae, providing an additional thermal buffer against the cold. They then passively drift with currents to other, sandier, sites, where they spend the remainder of the winter, surfacing for air every three or four hours. They remain dormant until March, unless they are disturbed or forced to come up for air.

Historically, these dormant turtles were selectively hunted on clear windless days when hunters could spot their carapaces on the shoal bottoms at low (neap) tide. The hunters preferred to find them where the water was four to eight meters deep, fashioning harpoon mainshafts seven to ten meters long to spear them. Two long-shafted harpoons were typically required to gain enough purchase to pull the turtle up from the bottom sediments, through the currents on the shoals, and into the boat. This effort took considerable prowess as well as carefully honed observational skills, since the turtles are not at all easy to see. Their carapaces are partially covered by several species of marine algae and by a sandy film, so that sometimes only the marginal scales are visible. The more the carapaces are covered with algal mats of Gracilaria textorii or Padina durvillaei, the longer the turtles have been dormant.

The Comcáac note that these dormant turtles are typically skinny from not eating. Their weight declines over the dormant season, but the flavor of their meat does not necessarily change. The ones that have retained any fat at all, it is said, have come out of dormancy early to forage in late winter in nearby eelgrass beds. According to Guadalupe López, when they do emerge from dormancy in late winter or in the spring, Bumphead Parrotfish (*Scarus perrico*) seek them out and graze the algae off their carapaces. A small crab called *moosni ihz*, "sea turtle's pet," is also sometimes found feeding on these algae. Neither crustacean expert Rick Brusca nor sea turtle expert Jeff Seminoff could suggest what crab this might be.

By the mid-1970s, this resident turtle population had become scarce in the central Sea of Cortés: once non-Indian hunters discovered the ease with which winter-dormant turtles could be hunted by diving with hookah gear, they quickly depleted most *iime* of year-round residents. While Jesús Rojo claims that Seri boatmen still encounter a few each year, Guadalupe López laments that most have left after being disturbed by the noise of motorized boats. A 1976 report by Felger, Cliffton, and Regal of winter-dormant sea turtles in the Sea of Cortés has led to similar reports of nonmigratory populations from other parts of the world, and to renewed conservation monitoring aimed at protecting these special populations.

Eretmochelys imbricata
English common name: Hawksbill
Spanish names: *carey, perico*
Comcáac names: *moosni quipáacalc, moosni sipoj*

The term *quipáacalc* refers to the uneven or overlapping plates of the first form of this species, whereas *sipoj*, "osprey," likens the second form's bill to the beak of the common fish-eating bird of prey of the Sonoran coast (in a manner reminiscent of both the English common name and the Spanish term *perico*, "parrot"). These two forms of Hawksbills, both of which are known for their pointed bills (more narrow in the latter form), their overlapping carapace plates, their single flipper claw, and their mild but unexceptional taste, are considered by the Comcáac to be closely related. *Moosni sipoj* is said to have a narrower head than the *quipáacalc* form, a black rather than a reddish brown carapace, and a white rather than a yellowish plastron.

Ramón Perales says both will rest along the bottom of the Canal del Infiernillo as Green Sea Turtles do, but usually pass by rather than spending much time in residence. They are more typically solitary travelers in the high seas. He says that they achieve weights of 80 to 130 kilograms at best, although the most recent exemplar was an immature one weighing only 18 kilograms, rescued from fishermen's nets.

Tortoise-shell craft items (which are made only from Hawksbill carapaces) have been traded since colonial times; these include hair ornaments, animal silhouettes, and finger rings (Griffen 1959). Today such items are made only when dead Hawksbills are retrieved from fishermen's nets, and are usually kept for personal use rather than being sold as a tourist craft. Cleotilde Morales has shown me rings, made by her and her husband from Hawksbill shell, which they keep for personal adornment. While some say that Hawksbills are seldom eaten because of their inferior taste, Ramón Perales says that the flavor of the Hawksbill is fine; it is simply different from that of Green Sea Turtles. Hawksbills are seldom intentionally hunted by the contemporary Comcáac.

Amalia Astorga told me a story about her grandson Esteban, then age five, who early one morning found a Hawksbill entangled in algae and cordage in the shallows in front of Desemboque. He brought his parents and grandparents to it in time for them to free it. He offered the Hawksbill some of his own saliva so that it could recover its wetness, and his father carved the boy's name in its shell for good luck before releasing it. This traditional ritual is infrequently practiced today.

Although recent reports by biologists state that Hawksbills are extremely scarce in the Sea of Cortés, certain Seri fishermen regularly encounter them and rescue them from nets. An immature Hawksbill was seen by Hugh Govan close to shore at the northwestern end of Bahía Kino in May 1999; I have found shell fragments of this species at both Tecomate and Campo Egipto. The Comcáac have no recollection of encountering Hawksbill eggs or nests near their camps, and no substantive documentation has been made of Hawksbill mating or nesting in the Sea of Cortés.

Lepidochelys olivacea
English common name: Pacific (Olive) Ridley
Spanish name: *golfina*
Comcáac name: *moosni otác*

The name *moosni otác* literally means "toadish turtle," for the generally flat, broadly convex body shape of Pacific Ridley Turtles reminds the Comcáac of the crouching form of the Spadefoot Toad (*Scaphiopus couchii*), which they call *otác*. A mature Pacific Ridley is the smallest of all adult sea turtles arriving along the Sonoran coast, sometimes weighing only 20 to 30 kilograms. The carapace is olive green in color, narrow and flared with four scutes at the front and broadening at the sides. The Comcáac say that the color of a Pacific Ridley plastron is yellowish. Its head is small, and it has only one claw per flipper.

Pacific Ridleys are said to undulate up and down rather than swimming on a level plane. They are only occasionally eaten because of inferior taste and small size.

Jesús Rojo claims that the Pacific Ridley is more a turtle of the high seas than of the Canal or of coastal lagoons, yet he recalls seeing many near Isla San Esteban in the summer. Although Felger and Moser (1985) claim that the *golfina* is now scarce compared to its former abundance, it is still frequently seen on the Sonoran coast, and isolated nesting continues throughout the Sea of Cortés, in the 1990s occurring as far north as Puerto Peñasco.

The times when hatchlings were found by contemporary Seri remain well remembered and widely discussed. Pacific Ridley babies were formerly scooped up from the beach sand and kept as pets, but they were sometimes so fragile that they died in captivity. Even the adults were easy to catch; hunters could pull them up into the boat without using harpoons, the turtles were so curious.

DERMOCHELYIDAE: Leatherback Turtles

Dermochelys coriacea
English common name: Leatherback
Spanish names: *siete filos, baúla, laúd*
Comcáac names: *moosnípol, xica cmotómanoj*

The name moosnípol, a contraction of moosni and coopol, "black," refers to the black or charcoal-green, rubbery, ribbed carapace of this animal, the largest marine reptile in the world as well as the largest and most sacred animal in the Comcáac bestiary. The Comcáac know it as an animal that prefers the high seas and reaches a length of 2 to 2.8 meters and a weight of up to 800 kilograms. They distinguish the Leatherback from other turtles not only by its oval, seven-rayed carapace, the lack of claws on its flippers, and its large head with blotched skin, but also by its capacity to shed tears, which they say it does to grieve the loss of a child.

An ancient beach encampment on the northeast side of Isla Tiburón is called *Moosnípol Quipcö,* "Thick Leatherback"; it is located where cliffs and mangrove fringe come together with the sand spit known as Punta Mala. According to Burckhalter (1999, 9), "The serrated mountain scarps that run the length of this imposing formation from south to north were said in one Seri myth to be the shell ridges of a giant leatherback turtle that rose up from the depths to form the sacred island."

The Leatherback is considered, along with boojums and Teddy Bear Chollas, to be one of the original kinds of humans or sentient beings, and so is deeply revered by all Comcáac. During the fiesta held on its behalf, the name moosnípol should not be used in its presence; rather, it must be spoken of indirectly as one of the xica cmotómanoj, meaning "weak things" or "fragile ones." This fiesta occurs when a Leatherback detours into the Canal from its normal foraging grounds in the high seas, where Guadalupe López says it eats jellyfish and the ctenophore known as *copsíij*, "sea gooseberry," or in Spanish as *agua mala*. There is a song about a Leatherback getting stung by sea gooseberries while feeding on them. Leatherbacks typically arrive in Comcáac territory between March and May. When a Leatherback hears songs sung for it in beach camps during the new moon, it is said to veer off course to renew contact with the Comcáac. Female Leatherbacks are said to come into the shallows on their own accord and then beach themselves; they may also be helped ashore by singing women. Historically, they were usually harpooned in the flipper, roped, and then guided or pulled into shore.

As soon as a Leatherback is found on the beach or in the shallows offshore, everyone present must abandon their other activities and prepare a fiesta for it; failing to do so might bring on death or the destruction of the entire community. A bower of ocotillo (*Fouquieria splendens*), adorned with branches of *xoop inl*, "elephant tree's fingers" (*Bursera hindsiana*) and other sacred plants, is constructed above the beached turtle, and its flippers, head, and carapace are painted with traditional designs (described at length in Espinosa-Reyna 1997). Everyone present, including non-Seri, must consent to be painted with the same designs, or they are asked to depart from the sight of the Leatherback.

Those who capture the Leatherback must provide drinking water to the turtle and to visitors, and they are also responsible for maintaining the bower or shade house, where the singing of sacred Leatherback songs occurs. Traditional songs, dances, and games also play into the ceremony, which many claim is the most sacred event in a Seri individual's life.

The fiesta is meant to pacify the Leatherback and to bring good luck to the community. After four days and one full night of veneration, the turtle is released back into the water, as David Burckhalter (1999) saw Guadalupe López do at Dos Amigos near Punta Sargento in April 1981. At that ceremony another Leatherback, which had been netted and was already weakened and injured when it was brought ashore, had died. On the rare occasion when such an animal does die, it is said to offer its body to the community for a ritual communion. Consumption of Leatherback meat outside this context is strictly taboo;

in fact, some say that even when it was ritually eaten, the ancestors had to close their eyes or they would go blind. The communion meal was said to taste of the "greens" that the Leatherback had last eaten. If a Seri individual is bluntly asked about the flavor of Leatherback meat, however, he will answer that it either is not eaten or has no flavor.

According to Jesús Rojo, the fiesta, if not the animal itself, induces visions of lasting importance among the participants. Guadalupe López added that Leatherbacks can understand what the Comcáac say in their presence and respond to them by coming and speaking in Cmique Iitom in their dreams. Formerly, the organizers who prepared the fiesta received a fee for their effort from their neighbors, but these days no one pays, and a wider diversity of families contributes to the effort. However, the fiesta is celebrated less and less frequently as fewer Leatherbacks arrive on the Sonoran coast. While there has been a large drop in Leatherback populations generally, the most dramatic decline has been in the number of females nesting in western Mexico, from more than 70,000 in 1980 to less than 1,000 in 1999 (Cliffton 1990; Sarti et al. 1996). Sandy Lanham of Environmental Flying Services flew the entire Pacific coast from Sonora to Panama in late winter of 1999 and found that counts of nesting females continue to decline.

According to Eckert (1999), this decline has not been caused by human predation on eggs or on gravid females on Mexico's nesting beaches, but by incidental take of Leatherbacks off the coast of Chile and Peru in a ten thousand–fold increase in commercial fishing efforts to obtain swordfish. While Mexico once had more Leatherbacks arriving on its beaches than any other country in the world, today even the sighting of a Leatherback carapace fragment or bone washed up on the Sonoran coast is cause for celebration among the Comcáac. Only two live Leatherbacks were seen near the Comcáac villages in the past few years, one in April 1999, which was released from nets and sent on its way by Seri fishermen, the second in September 1999, seen by Tad Pfister south of Playa San Nicolás; it was found dead a month later.

Leatherbacks are known by the Comcáac as the only species of sea turtle that comes into the Canal del Infiernillo without necessarily being accompanied by others of its kind, and the individuals that do so are small—300 to 400 kilograms, as opposed to the 500- to 800-kilogram individuals once seen in the high seas. Guadalupe López and Ramón Perales say that Leatherbacks are completely absent from the Sea of Cortés in the winter. When they formerly arrived in spring, they were seen coming northward across the high seas. If they were seen with other turtles, it was with *cooyam*, the young migratory Green Sea Turtles. Now, both species' migrations are disrupted by gill nets, long-lines, and drift nets.

The Comcáac are deeply concerned that so few of their youths have seen live Leatherbacks or are familiar with the fiesta. They have therefore signed a document requesting permission to visit Leatherback nesting beaches in Central America to perform their ceremony there and pleading for stronger international conservation efforts on behalf of the *xica cmotómanoj*. They believe that because all Leatherbacks can understand Cmique Iitom, if they sing songs to the turtles where they come to nest, those present might be better protected.

Dermochelys coriacea × *Caretta caretta*
English common name: Leatherback-Loggerhead hybrid
Spanish name: *siete filos híbrido con carey*
Comcáac name: *moosnípol xpeyo (quimozíti)*

In May 1999 Alfredo López described a hybrid sea turtle to Jeff Seminoff and me, of a shape and coloration that he had seen only once in his life. At Campo Víboras on Isla Ángel de la Guarda, Alfredo tried pulling into his boat a sea turtle with a gray-green carapace that was flattened in shape like that of a Pacific Ridley's but had characteristics of Leatherback and Loggerhead carapaces; the animal's head also reminded him of the latter two species. He said that it bit him with its many small teeth and showed us the scars on two fingers that this unique turtle had left as a *recuerdo*. Stunned by its singular beauty, he released it and never saw it again. The term *quimozíti*, used descriptively, means "to be half one thing, and half another" (Moser, Moser, and Marlett, in prep.).

LIZARDS

CROTAPHYTIDAE: Collared and Leopard Lizards

Crotaphytus dickersonae and *C. nebrius*
English common names: Tiburón/Dickerson's Collared Lizard and Sonoran Collared Lizard, respectively
Spanish names: *lagartija de collar, escorpión de la piedra*
Comcáac name: *hast coof*

Hast coof means "chuckwalla-like lizard of rocky habitats." These medium-sized lizards frequent rocky uplands, with *C. nebrius* occupying the mainland and *C. dickersonae* accessible to the Comcáac only on Isla Tiburón. It is considered a *coof*-like resident of mountain ranges and will not even come down to flatter, sandier terrain, some Seri say.

Most of the Seri whom I have interviewed readily identified *hast coof* from photos or from preserved specimens. It is known by its behavior as much as it is by its markings and its habitat preferences. Collared lizards are considered to be pugnaciously territorial. Nacho Barnet told me that "if you throw a rock at it, it will not run away from you, but will hold its ground or threaten to attack you." On Isla Tiburón in October 2001, a Comcáac crew of Desert Tortoise surveyors watched, teased, and threw pebbles at a brilliantly colored collared lizard for more than a half hour, with the lizard several times coming within one meter of its observers. While not considered to be dangerous to humans, the collared lizards of the Comcáac homelands are characterized as voracious predators, eating butterflies, mosquitoes, ants, bees, and flies.

Although it has not been killed or eaten by the Comcáac in recent times, it does figure in their lore. Angelita Torres is aware of several people who still know songs about *hast coof*.

Gambelia wislizenii
English common name: Long-nosed Leopard Lizard
Spanish names: *lagartija de leopardo, cachora*
Comcáac name: *hantpízl*

Hantpízl may be derived from *hant*, "earth, soil, land," and *cahpízl*, "ridged, arched, furrowed." Nearly every Seri individual who sees this colorfully spotted lizard recognizes it as *hantpízl*, a dweller of the rocky uplands. It is capable of darkening or lightening its skin color, and can shift to a wide variety of prey items as needed. Leopard Lizards are widely distributed geographically, from northeastern Baja California northward and into coastal and central Sonora. They sun themselves on boulders and outcrops, staying active even on the hottest of days. The males are larger and can be distinguished from the females by their swollen femoral ridges.

The Comcáac claim that *hantpízl* consumes flies, ants, fruits, and even other lizards. Some say they have seen it put other lizards' heads in its mouth, sucking out their tongues. *Hantpízl* is a wrestler and a biter. Malkin (1962) was told that the bite of Leopard Lizards was toxic and potentially fatal only to Seri victims, not to other animals or other kinds of people.

IGUANIDAE: Iguanas and Chuckwallas

Ctenosaura conspicuosa and *C. nolascensis*
English common names: Isla San Esteban Spiny-tailed Iguana and Isla San Pedro
    Nolasco Spiny-tailed Iguana, respectively
Spanish name: *iguana*
Comcáac name: *heepni*

Not all Seri individuals immediately know what an iguana is called in Cmique Iitom, for it is seldom seen today except by those men who fish or hunt out on Islas San Esteban and Cholludo. Nevertheless, more than 85 percent of the adults shown photos of this iguana distinguished it from chuckwallas by its spiny crest and tail and by the greenish skin of its hatchlings. Biologists know this iguanid by its unmistakable shape, size, spininess, and habit of climbing high into columnar cacti and trees. Older men were aware of its presence on Isla San Pedro Nolasco, which they called Hast Heepni Iti Ihom, and some knew that it regularly occurs on the coast south of Guaymas. Chapo Barnet claimed to have seen one once in a thicket near Bahía Kino while he was gathering wood there, but there are no other oral histories or documented specimens of free-living animals north of Guaymas.

Borys Malkin (1962), who thought only *Dipsosaurus* and *Sauromalus* occurred on Isla San Esteban naturally, suspected that the Moreno family was biologically ill informed when they told him of another large iguanid on the island. Herpetologist Lee Grismer agrees in the sense that San Esteban is too far away from source populations of *Ctenosaura* for it to have been naturally dispersed there; cultural dispersal from Hast Heepni Iti Ihom is more plausible.

The males are known by the Comcáac to have enlarged crusted pores on their femoral ridges. The Comcáac are also aware that this iguanid lizard is herbivorous, observing that it eats the flowers of ocotillos, cacti, and seasonal herbs, as well as ironwood foliage.

Some claim that the island iguanas are so tame that they do not bite people, although many biologists and a few Seri have been bitten when handling them. Other Seri, such as José Juan Moreno, claim that spiny-tailed iguanas are "más cabrón" and harder to catch than chuckwallas, but relish both species as food.

As Darwin (1997) quipped about other large iguanids, "The meat of these animals when cooked is white, and by those whose stomachs rise above all prejudices, it is relished as very good food. Humboldt has remarked that . . . all lizards which inhabit dry regions are esteemed delicacies for the table." A number of Seri individuals over fifty years of age have eaten spiny-tailed iguana, preferring the shoulder and back meat. A hand-caught iguana was ritually killed by reciting a blessing over it, whipping it in a circle overhead, then hitting its head against a boulder. Although some Seri are concerned that iguana numbers have declined since the introduction of rats, they assert that their own hunting of the animal did not leave spiny-tailed iguanas scarce on Isla San Esteban. They claim that their

Comcáac forefathers moved this species to Isla Cholludo to establish a food supply there in case fishermen became stranded during rough weather.

Dipsosaurus dorsalis
English common name: Desert Iguana
Spanish name: *porohui*
Comcáac names: *meyo, ziix tocázni heme imocómjc*

One unanalyzable name for this modest-sized iguana, *meyo*, is known only to a few elderly Seri individuals, whereas everyone else calls Desert Iguanas by the name ziix *tocázni heme imocómjc*, which can be loosely translated as "thing that, if it bites, prevents from arriving to home camp." This descriptive phrase warns of the legendary harm that a Desert Iguana—or a man named Meyo—once inflicted on an ancestor of the Comcáac. In ancient times, a beloved man announced to his family that he would walk to see relatives in another camp, but he never arrived. When his body was at last found, a *meyo* was seen moving away from the hand of the corpse, as if the man had been carrying it just prior to being killed. It was forever after called ziix *tocázni heme imocómjc* to warn other Seri individuals away, lest they suffer a similar fate.

In fact, this lizard is considered to be so psychologically dangerous as well as poisonous that my queries about it were largely rebuffed. Its ghostly paleness reminds some biologists of geckoes; this trait may contribute to the Comcáac aversion to both animals. Felger (1990) was told that *meyo* was assiduously avoided because of its toxicity, and Ernesto Molina told me that its bite could kill a person "on the spot." This peril, however, appears to be culture-specific. The late Antonio López was aware the O'odham ate this iguana, but no other Seri would comment on its edibility. López also knew of a song about *meyo*.

Desert Iguanas are common up in the Sierra Comcáac and on several of the islands close to Baja California. Malkin's notes (1962) that the Comcáac consider it to be herbivorous, arboreal, egg laying, sexually dimorphic, and edible may in fact refer to the spiny-tailed iguana and not to the Desert Iguana.

Sauromalus hispidus, S. obesus (formerly *S. ater*), *S. varius*, and hybrids

English common names: Spiny (Black) Chuckwalla, Northern Chuckwalla, and
   Piebald (Isla San Esteban) Chuckwalla, respectively

Spanish name: *iguana*

Comcáac names: *coof coopol, ziix hast iizx ano coom,* and *coof,* respectively

Coof is related to the verb *coof,* meaning "to hiss, snort, or puff" like a chuckwalla.
Ziix hast iizx ano coom loosely means "living thing that wedges itself into a rock
crevice." Becky Moser informs me that *iizx* is from the verb *cazx,* "to tear or to
split," which describes the sound heard when one tries to remove an inflated
chuckwalla from its crevice. These three chuckwallas are distinguished from one
another by color and size, with the *coof coopol* of Islas San Lorenzo, Ángel de la
Guarda, and nearby islets being the darkest and least blotchy in coloration. The
shrubs, caves, and crevices that these species frequent are well known to contemporary Comcáac hunters, who formerly scared the animals out of hiding by
whirring wandlike branches of *Asclepias* milkweeds overhead to make a windlike
noise (Felger and Moser 1985). The bloated animals were also deflated with
pointed sticks and removed by hand. Although the San Esteban people had few
land mammals available as food, it is said that they did not particularly like the
taste of chuckwallas so did not overexploit them (Bowen 2000). In one cave
near the mouth of Arroyo Limantur, says Ernesto Molina, during the winter they
stacked themselves atop one another clear to the ceiling, in a behavior which he
called *ano yasnan,* "blanketing" or "stacking"; here they would have been particularly vulnerable to overharvesting.

   Tom Bowen and the Mosers have recorded an oral history which suggests
supernatural punishment for those who senselessly kill or drown chuckwallas.
As Bowen (2000) puts it, "*Hast Cmique* said it is wrong to kill animals without
reason. . . . A person who mistreats a chuckwalla by throwing it into the sea
will be punished when he is at sea, perhaps by being subjected to strong winds."

   The chuckwallas themselves are known to relish ironwood leaves and cholla
cactus flower buds. A song composed by Jesús Félix tells of a chuckwalla waiting for the sun to break through the fog before it goes to eat its favorite things.

When José Juan Moreno went foraging with me on Isla San Esteban, he found a female chuckwalla on a ridge covered by cacti and agaves, and caught it with little trouble. At first I thought her snout had blood on it from biting José Juan's finger, but he laughed and said that her reddish mouth was from subsisting on *Pitahaya agria*, "sour cactus fruit" (*Stenocereus gummosus*), over the previous weeks. Moreno disabled his catch by twisting her legs, and kept the animal alive until he was back home in Punta Chueca. He savored the meat, including the fat-laden tail, the next morning.

Zooarcheologist Richard White (pers. comm.) informs me that the Comcáac have occasionally brought island chuckwallas to the mainland for release. Only recently have they formally initiated translocation for the purpose of wildlife conservation. Several Comcáac reported to me that their ancestors released chuckwallas from San Esteban onto Tiburón and Punto Sargento, and I was able to confirm the latter. In September 1998, the Comcáac community in Punta Chueca launched a chuckwalla captive breeding program with one male and three females from Isla San Esteban. By September 1999, the weakest female had died, but three other males had been recruited from the island. The Seri then discussed plans to shift the sex ratio of the group and expand their foraging area and native plant diet. The project, which they named *Ziix Hast Iizx Ano Coom Hant An Ihacámot*, meaning "Chuckwalla Nursery Grounds Refuge," remains successful as of the date of this publication.

PHRYNOSOMATIDAE: Sand, Rock, Spiny, Horned, Brush, and Side-blotched Lizards

Callisaurus draconoides
English common name: Zebra-tailed Lizard
Spanish name: *perrito*
Comcáac names: *ctamófi, ctoicj, matpayóo*

This slender, elongated lizard is superficially similar to Lesser and Greater Earless Lizards, but its behavior and habitat preferences are distinctive. It is recognized by the way it wags its striped tail in a scorpionlike motion and by its high-speed run on its back legs. These movements, more than the animal's color or shape, are what brings the name *ctamófi* to the lips of a Seri seeing one of them. Once, when beaching a panga at Ensenada del Perro on Isla Tiburón, we saw several Zebra-tailed Lizards running upslope ahead of us. Despite seeing them for but a split-second out of the corner of their eyes, my Seri companions said "ctamófi" matter-of-factly. And yet when Rosenberg (1997) showed photos of this lizard to Comcáac people under forty, none used this name, and only 59 percent of Comcáac adults over forty used it. Even when I have placed a pre-served specimen of this lizard in people's hands, people did not call it *ctamófi*; instead they called it *ctoicj*, an archaic name, or *matpayóo*, a name that refers only to the juveniles. Others simply call it *haquímet*, the generalized term for small lizards. As Guadalupe López reminded me, "Traditionally, the animals of the desert scrub had two, maybe three names."

Malkin (1962) was told by Seri individuals that male Zebra-tailed Lizards were larger than females and that females cared for both eggs and hatchlings. *Ctamófi* are most frequently encountered on beaches or in other sandy areas. If caught, the Comcáac say, *ctamófi* will bite, but it is nonvenomous. However, Guadalupe López said that "it has its power. . . . If you become its enemy, it can inject some-thing into you that makes your eyes red."

Malkin's report (1962) that it was once eaten during periods of food scarcity has not been confirmed. In one song, this lizard talks to the people, telling the Comcáac how they should bury themselves in the winter to escape the cold.

Phrynosoma solare
English common name: Regal Horned Lizard
Spanish name: *camaleón*
Comcáac name: *hant coáaxoj*

Hant coáaxoj may simply mean "earthlike lizard." There is no mistaking a horned lizard for any other reptile, although there is little recognition among the Comcáac today that two or more species may occur within their historic range (Sherbrooke 1981). The Regal Horned Lizard is the one species encountered with any regularity today, and it is among the most favorite animals with which Comcáac youths have contact (Rosenberg 1997). They find horned lizards in sandy areas or near ant mounds, and often bring these lizards home as pets. Some children keep them only a few days, then release them where they were found. However, at least one Seri boy raised a hatchling to maturity—a feat achieved by few herpetologists (Sherbrooke 1987)—taking it out to feed on ants. Black ants are preferred, but flies and grasshoppers are said to be eaten as well. Malkin (1962) was told that horned lizards are viviparous and also have scrotums.

Ernesto Molina explained to me that if molested, hant coáaxoj will shoot a ray-like stream of fluid into the joints of its assailant—into the knees, elbows, or shoulders. Some Seri believe that this fluid is stored in the spines crowning its head. The resulting malady is much like the "staying sickness" that afflicts neighboring O'odham who have broken a taboo associated with this animal (Bahr et al. 1979). If a Seri individual accidentally disturbs a horned lizard, his or her companions must immediately chant "Haxáx cöquépe!"—Ay, let us comfort you!—to prevent the animal from becoming offended or frightened. Children still observe this ritual when out playing. According to Rosenberg (1997), 83 percent of the Comcáac she interviewed over forty years of age believed that horned lizards shoot this fluid into molesters, whereas only 41 percent under forty believed it.

Nearly everyone, however, knew the nursery rhyme about a horned lizard that was bitten by ants and died while gathering ironwood kindling in the mountains. The song is really about a man named Hant Coáaxoj who lived in mythic times; his death on their behalf is one reason the Comcáac now protect horned lizards. In prefacing the song, singers often exhibit an affection and respect for horned lizards that is usually reserved for family members. Another song fondly and humorously describes the animal's bumpy back and head. The Comcáac demonstrated the same affection for other horned lizards when I showed them live animals of other Sonoran Desert species.

Sceloporus clarkii and *S. magister*
English common names: Clark's Spiny Lizard and Desert Spiny Lizard, respectively
Spanish names: *cachorón* (both species) and *bejori* (*S. magister*)
Comcáac name: *haasj*

Haasj (pl. *haaslca*) is an unanalyzable primary lexeme, and its use by the Comcáac to refer to these lizards suggests a long association with them. Although people readily identify all spiny lizards as *haaslca*, they commonly see the larger Desert Spiny Lizard in all habitats near their villages, whereas Clark's Spiny Lizard tends to be restricted to canyon habitats. Both species are known from Isla Tiburón, but their overall geographic ranges are not well known to the Comcáac today. Spiny lizards, they say, prefer desert scrub valleys to mountainous habitats, and they will climb walls and trees. They are known to stay underground for long periods, perhaps as long as an entire year. When they emerge, they eat flies, flowers, and herbs. The males are known to be territorial fighters; their skin can get torn from such fights, and the characteristic texture of their shredded or molted skin can be identified to these species by the Comcáac.

Desert Spiny Lizards are occasionally caught on strings by boys, who keep them for several days (Malkin 1962). Although not eaten or otherwise used economically, spiny lizards are subjects of several Comcáac songs and lewd jokes referring to the "push-ups" the male *cachorones* perform. In Sonoran Spanish slang, the verb *cachorrear* means "to engage in sexual foreplay." In neighboring O'odham rancherias, women were afraid of having to urinate behind a woodpile inhabited by spiny lizards, for fear that lizards would run up their skirts and lodge themselves (painfully) in the private parts.

Urosaurus ornatus
English common name: Tree Lizard
Spanish names: *cachora matorralera, lagartija*
Comcáac name: *hehe iti cooscl*

Hehe iti cooscl means "speckled gray thing high up in a tree." This inconspicuous lizard is identified as much by its habitat preferences as by any other trait. It is known to climb trees, shrubs, and cacti, and is sometimes confused when caught by hand with other small gray and brown lizards.

Uta stansburiana
English common name: Side-blotched Lizard
Spanish names: *cachora gris, lagartija de cercos*
Comcáac names: *tozípla, yax quiip, matpayóo, ctoicj*

Tozípla is an unanalyzable primary lexeme; the less commonly used name *ixax quiip* may refer to the lizard's straight snout. *Matpayóo* is a term used for the young, while *yax quiip* and *ctoicj* may be archaic names for both adults and young. While Comcáac observers readily recognize the side blotches on this species, they still sometimes confuse these animals with Tree Lizards or with Zebra-tailed Lizards. Both Comcáac names may, in fact, on occasion be used for other small, brownish lizards. There is a song about a *tozípla* living in a cave, where arrows (or thorns?) pierced its body like so many white flowers (Rosenberg 1997). *Tozípla* appears to be the kind of lizard that Amalia Astorga named Efraín and fed for seven years before his untimely death (Astorga, Nabhan, and Miller 2001). His progeny continued to visit Amalia for some time.

EUBLEPHARIDAE: Eyelidded Geckoes

Coleonyx variegatus
English common name: Western Banded Gecko

Spanish names: *geco, salamanquesa de franjas*
Comcáac name: *cozíxoj*

Banded geckoes, the most frequently encountered of the geckoes on the Sonoran mainland, are considered the most dreadful example of a *cozíxoj* and are assiduously avoided. Although all Seri individuals identified this gecko as a kind of *cozíxoj* when I have shown them preserved specimens, not all believe it to be as dangerous as elders' stories claim. Janice Rosenberg (1997), for example, found that only 34 percent of youths between nine and nineteen years of age believe the stories that its toxic saliva causes rashes, that it can harm unborn babies if pregnant mothers see it, or that it causes lung illnesses and the deterioration of human flesh. Punta Chueca residents have told me that it can injure both men and women by squirting blood on them, and that difficult deliveries or stillbirths may be suffered by women who molest banded geckoes. Nacho Barnet said that Comcáac children are afraid to dig in the sand where geckoes have been found on beaches. If one lies down where a gecko is buried, its silhouette will be burned into the victim's flesh. Ironically, they are not considered dangerous to other animals; Malkin (1962) learned from the Comcáac that leopard lizards will eat banded geckoes.

GEKKONIDAE: Geckoes

Phyllodactylus xanti
English common name: Peninsular Leaf-toed Gecko
Spanish names: *geco, salamanquesa*
Comcáac name: *cozíxoj*

Several kinds of leaf-toed geckoes inhabit the islands of the Sea of Cortés and have been reported at a few coastal locations on the Sonoran mainland. The Seri I have interviewed about leaf-toed geckoes are not particularly conversant with the traits that distinguish them from banded geckoes. When shown living geckoes during a visit to the Arizona-Sonora Desert Museum, Seri men regarded them all as members of the same *cozíxoj* family. Their distribution is poorly known.

TEIIDAE: Whiptail Lizards

> *Cnemidophorus burti* and *C. tigris*
> English common names: Canyon Spotted Whiptail and Western Whiptail, respectively
> Spanish name: *güico*
> Comcáac names: *ctoixa, ctoixa iipíil* (hatchlings)

Although adult whiptails are called simply *ctoixa*—an unanalyzable primary lexeme—the hatchlings are so brilliantly blue in hue that they are given the name *ctoixa iipíil*, "indigo whiptails." All whiptails are fast, sleek lizards that the Comcáac say move unlike any others. The Comcáac most frequently encounter Western Whiptails in sandy desert scrub in and near their villages, but Craig Ivanyi and I saw mostly Canyon Spotted Whiptails in comparable habitats on Isla Tiburón's beaches and washes. Both species occur on the island, where their multicolored spots overwhelm any striping visible in mainland populations. Several Seri have told me, too, that hatchlings of both species are more intensely blue on Isla Tiburón than on the mainland.

Malkin (1962) was told by his Seri companions that male whiptails were larger and had more-swollen femoral ridges than females. His identification of whiptails is doubtful, however. He witnessed Comcáac children catching them with strings and keeping them as temporary pets before killing them, a testament to the youngsters' skill. In fact, catching whiptails with nooses is challenging even for most field herpetologists.

In an allegorical nursery song described to me by the late Eva López, a mother whiptail warned her children not to go out on a particular trail, for there was a snake there that had another lizard in its mouth and would also try to eat them. One child bragged that "no one is going to eat me," then disobeyed her. It was eaten!

XANTUSIIDAE: Night Lizards

Xantusia vigilis
English common name: Desert Night Lizard
Spanish name: *cuija, salamanquesa*
Comcáac name: *cozíxoj*

These nocturnally active lizards typically rest under rocks, tree trunks, or cactus carcasses. Possibly near their southernmost limits in Comcáac territory, they are not generally well known in Sonora (Felger 1965). Although several of the Seri men and women I have talked with admit that they know there are different lizards subsumed under the name *cozíxoj*, they consider all of them repugnant and potentially dangerous. While some Seri will not even touch them, others will do anything to evict them from their campsite or household. Touching them causes psychosomatic illnesses of mythic severity.

HELODERMATIDAE: Gila Monsters

Heloderma suspectum
English common name: Gila Monster
Spanish name: *escorpión pintado*
Comcáac name: *paaza*

Of all the terrestrial reptiles in the Comcáac homelands, *paaza*—its name an ancient, unanalyzable primary lexeme—is perhaps the most distinctive. Quickly identified from photos, drawings, and specimens, it is among the least preferred species in the local menagerie (Rosenberg 1997). The Comcáac, who can imitate its slow lumbering gait, say *paaza* is active both day and night. Although not afraid of encountering a *paaza* under normal circumstances, they are aware that it can be "muy cabrón" and fight off potential predators.

Few biologists I know feel that they can accurately sex Gila Monsters, the size, shape, and color of males and females being much the same, but certain Seri individuals claim to identify the gender of these animals by head and body width. The Comcáac encounter Gila Monsters in a variety of habitats, from rocky outcrops to loose sandy washes and desert scrub flats. *Paaza* are known to hibernate in the banks of arroyos or in small caves, sometimes in the presence of Desert Tortoises or rattlesnakes. During part of their active season, *paaza* may be tending four to eight eggs or feeding on beetles and other ground-dwelling invertebrates.

Ernesto Molina has told me that when angered, a *paaza* will spit or squirt a substance from its mouth four meters or more. This substance is not considered toxic, but is nevertheless foul smelling and capable of causing rashes. The animal's bite, however, is known to be toxic, though it has seldom, if ever, resulted in the death of a Seri individual.

In the past, Gila Monster tails were eaten when other food sources were in scarce supply, being fried in oil (Malkin 1962). They have also been used medicinally, to treat psychosomatic illnesses. Gila Monster skin is sometimes heated and placed on a person's forehead to cure a headache. A few villagers use the beaded skin from its tail to make rings, necklaces, wristbands, and hat bands. Juana Herrera still makes rings and necklaces of Gila Monster skin for good luck.

SNAKES

BOIDAE: Pythons and Boas

> *Boa constrictor*
> English common name: Boa Constrictor
> Spanish names: *boa, corúa*
> Comcáac name: *xasáacoj*

Boa constrictors are known as the largest nonvenomous snakes that live near water holes in the mountains of northwest Mexico. They, along with the mythic *corúa*, take their name of *xasáacoj* (which may mean "chewed fruit" [Felger and Moser 1975]) from the prostrate, snaking cactus *Stenocereus alamosensis*, which they are said to resemble in both shape and girth. The cactus and the constrictor reach their northern limits in Comcáac homelands, while the mythic *corúas* protect springs as far north as the Hopi mesas. Curiously, the overall distribution of Boa Constrictors in Sonora, and especially in the southern portion of Comcáac homelands, is about the same as that of the cactus, which reaches into Seriland as far north as Siete Cerros, between Hermosillo and Calle 12, where it may have been introduced by the Comcáac centuries ago. For some reason, in their masterwork on Seri ethnobotany, Felger and Moser (1985) did not reiterate the metaphorical link between Boa Constrictors and this cactus.

Lowe (1959) provided the first intriguing documentation of Boa Constrictors occurring north and east of Guaymas, but there have been few recent sightings of them by biologists north of Bahía San Carlos (Felger 1999; Van Devender, pers. comm.). In October 1999, Laurie Monti and I had the good luck of being led by José Antonio "Delfino" Magayan to where a Boa Constrictor was eating a songbird at the top of a six-meter-tall *guaymúchilillo* (*Havardia sonorae*). Delfino claimed that *corúas* are uncommon in the Sierra El Aguaje, except around spring-fed waterholes or when they are camouflaged in the checker-barked canopies of *Havardia sonora* and *bebelama* trees (*Sideroxylon occidentale*). He said that they practice sit-and-wait predation, sometimes "hypnotizing" the birds that they then ambush as their prey. The Boa Constrictor had apparently caught the songbird just prior to our arrival, because we fleetingly witnessed feathers protruding from its mouth before the entire bird disappeared inside the snake.

Much of former Boa Constrictor habitat near the northern limits of its range has been converted to irrigation agriculture. However, the mythic *corúas* remain alive in the minds of many Sonorans. They are large and without rattles, just as Boa Constrictors are, although José Luis Blanco carved me a six-foot-long, three-inch-thick *corúa* that did have a small rattle. His neighbors told me that he was well aware that *corúas* lack rattles, but that he took artistic license while fashioning it out of *paij*, "driftwood."

Charina trivirgata
English common name: Rosy Boa
Spanish name: *boa rosada*
Comcáac names: *(ziix) haas ano cocázni, maxáa*

Ziix haas ano cocázni loosely means "thing living among mesquite like a rattlesnake," though it is sometimes called *maxáa* as well, as it was by Alfredo López when we caught one on Isla Tiburón in September 1998. This thick reddish brown snake is not as well known among the Comcáac as are others, nor is its range in coastal Sonora well known to biologists (Lowe 1959; Lawler, pers. comm.). Van Devender (pers. comm.) and Grismer (1999) have recently confirmed its presence on several of the small islands off the coast of Baja California, and in the Sierra Bacha north of Desemboque.

COLUBRIDAE: Colubrid Snakes

Arizona elegans
English common name: Glossy Snake
Spanish name: *culebra brillante*
Comcáac names: *coimaj quiimapxiij, cocaznáacöl*

Sometimes confused with the Gopher Snake by Seri and non-Seri alike, this nocturnal predator is distinguished from other *cocaznáacöl*, "large rattlesnake-like serpents," by its smooth, shiny scales. Its coloration is described as *quiimapxiij* by Alfredo López, that is, "brownish with many dark colors." The Glossy Snakes are found in Desemboque close to the beach, but they also are said to frequent coastal plains of desert scrub and dense vegetation in the mountains. Craig Ivanyi and I have found this snake among weedy amaranths in roadsides along abandoned fields north of Calle 12. Its range of local habitat use is said to be much like that of the Western Long-nosed Snake. The Comcáac know it to eat rats, mice, small squirrels, and lizards. Not considered venomous, it is occasionally killed to obtain bones for necklaces.

Chilomeniscus stramineus (incl. *C. cinctus*)
English common name: Bandless Sandsnake
Spanish names: *culebra de los médanos, coralillo falso*
Comcáac name: *hapéquet camízj*

Hapéquet camízj means something akin to "causing a pregnant woman to have a well-formed child." The bandless morph of this sand-loving snake is treated with great affection by the Comcáac. They know that it burrows effortlessly through sand, and if they encounter one while digging, they will hold it in their hands and stroke it, then release it. It will not bite them; for their part, the Comcáac feel that they must protect this friendly snake from predatory coralsnakes and uncaring humans.

To ensure that a baby will be born with lovely skin, this snake is captured and passed across the belly or the small of the back of the expectant mother. This ritual, performed less and less frequently, is also intended to give pregnant women hopeful feelings and help them give birth to good-looking children.

Chionactis occipitalis
English common name: Western Shovel-nosed Snake
Spanish name: *coralillo falso*
Comcáac names: *hant quip*

Hant quip combines *hant*, "earth, soil, sand," with the verb *quip*, "to dig, shovel, excavate"; it is thus not directly related to the term used to describe sand dunes, *hant quipcö*, "thick sand." Nevertheless, it is a remarkably apt name for this fossorial snake with a shovel-shaped snout, which it uses to burrow through dunes. This cream-and-brown-banded serpent also inhabits sandy coastal flats and beaches like those at Punta Chueca, where I live-captured a small one at dusk in March 1998. The young villagers I showed it to called it a false coralsnake in Spanish, and knew that it was nonvenomous. Others I have shown preserved specimens confused it with Banded Sandsnakes, which inhabit the same habitats; both are infrequently seen, since they are primarily nocturnal. Amalia Astorga and Adolfo Burgos knew that shovel-nosed snakes eat larvae and other invertebrates, and claimed that they occur on the northeast stretches of Isla Tiburón as well as on the Sonoran coast.

Hypsiglena torquata
English common name: Nightsnake
Spanish name: *culebra de la noche*
Comcáac name: *coimaj coospoj*

This prowler is known for its strictly nocturnal habits and for the large, dark blotches on its neck and the rich mottling along its body (*coospoj* means "blotched, spotted, colored"). Guadalupe López claims that Nightsnakes live on prey such as scorpions, larvae, and millipedes, which it finds in desert scrub. Craig Ivanyi and I have encountered it in sandy swales on the edge of the hypersaline estuary Laguna de La Cruz, where ancient cactus forests persist on higher ground. It also frequents paved roads at night. On Isla Tiburón, I captured a six-inch-long Nightsnake on the beach near Punta Palo Fierro in October 2001, not long after a hurricane had brought rain to the area.

Masticophis bilineatus; M. flagellum; M. slevini
English common names: Sonoran Whipsnake, Racer; Sonoran Coachwhip; Isla San Esteban Coachwhip, respectively
Spanish names: *culebra látigo; chirrionera, matorralera; alicante matorralera, alicante chicotera*, respectively
Comcáac names (for all three species): *coimaj coopol, coimaj cmasol, coimaj cquihjö, maxáa*

Coimaj and maxáa are unanalyzable primary lexemes used to refer to multiple species of nonvenomous snakes; the adjectives attached to *coimaj* are color descriptors.

Each time we have live-captured a Sonoran Coachwhip, for example, the Seri men with us first call it a *coimaj*, then add a color modifier to the name to describe its blackness or reddishness. Jesús Rojo claims that the whipsnake most commonly seen is called *maxáa*; he identified it from a field guide by pointing to the color drawing of *M. flagellum*, the longer Sonoran Coachwhip, which lacks the longitudinal stripes typical of other whipsnakes. However, there is great color and size variation within *M. flagellum*, which the Comcáac describe but do not formally label, nor do they regularly distinguish between the three scientifically recognized species in the genus *Masticophis*. Alfredo López knew, however, that there are closely related forms, and he claimed that they hybridize on Isla Tiburón. The Seri men who have seen the coachwhip on Isla San Esteban call it a *coimaj*, but consider it somewhat different in coloration from whipsnakes on the mainland. All three species have probably been killed to provide bones for necklaces.

The more common two species of these snakes are regularly seen both on Isla Tiburón and on the mainland. Cornelio Robles once killed a two-meter-long Sonoran Coachwhip with a blackish head and an orangish tail, in a canyon near Campo Vaporeta on the southwest side of Isla Tiburón. He and others frequently see these snakes climbing into trees, particularly mesquites.

In September 1998, Richard Nelson and I encountered an orangish-red Sonoran Coachwhip attempting to swallow a Tiger Rattlesnake on Isla Tiburón; when we returned to camp with the dead rattler and told Alfredo López what we had seen, he replied simply that this "*coimaj es más cabrón que cocázni.*" Biologists have determined that this fast-moving snake either actively pursues lizards and other prey by chasing them or "sits and waits" in the shade, conserving energy so it can ambush unsuspecting prey (Jones and Whitford 1989). It is likely that our coachwhip used the latter strategy on the rattler, in this case waiting in the shadows of an ironwood at the edge of a wash.

Jesús Rojo has been recorded singing about a coachwhip going into a crevice to find food, only to be attacked by bees and wasps. Even a swift, agile, crafty predator sometimes gets its just deserts.

Oxybelis aeneus
English common name: Brown Vinesnake
Spanish names: *guirotillo, bejuquillo*
Comcáac name: *hamísj catójoj*

This long, slender brownish snake with barklike skin is likened to limberbush stems by the Comcáac, as it is to vines by Spanish- and English-speaking observers. *Hamísj* is the Seri name for the limberbush, also known as *torote blanco* (*Jatropha cinerea*); *catójoj* means something like "to clump" or "camouflage." This echoes the lovely description that Van Devender, Lowe, and Lawler (1994, 37) use to describe Brown Vinesnakes in their habitat: "The wavering rigid body strongly resembles a slender branch moving in the wind."

The Comcáac claim that five to eight individuals of this arboreal predator will group together in the same tree, and that these snakes will seldom be seen on the ground. When mating, these snakes have been seen coiled up around one another like a ball. Herpetologists suggest that both their cryptic coloration and their slender body and tail not only protect them from predators but also allow them to ambush Tree Lizards, earless lizards, spiny-tailed iguanas, and birds that venture within their grasp (Van Devender et al. 1994). They are capable of injecting venom into their prey.

Phyllorhynchus decurtatus
English common name: Spotted Leaf-nosed Snake
Spanish names: *víbora de la noche, culebrita*
Comcáac name: *cocaznáacöl*

This secretive snake is not well known by the contemporary Comcáac. When Craig Ivanyi and I pointed out its distinctive rostral patch on one live-captured snake, the Seri individuals who handled it commented only that they knew it had no rattle. They suggested that it eats geckoes and larvae, leaving eggs and seeds alone. Craig and I found one of these nocturnal snakes alive on the highway between Bahía Kino and Calle Doce, amid abandoned fields being revegetated with desert scrub.

Pituophis melanoleucus
English common name: Gopher Snake
Spanish names: *cincuante, víbora sorda*
Comcáac name: *cocaznáacöl*

The most commonly seen nonvenomous brown-and-yellow snake in the Comcáac homeland is likened to large-sized rattlesnakes (*cocázni* contracted with *caacöl*), as its name, *cocaznáacöl*, suggests. The Comcáac have seen it raise its head and move its tail as if it were mimicking a rattlesnake; its name humorously connotes that it "thinks it is a rattlesnake." It is ubiquitous between Bahía Kino and Desemboque, and is common on Isla Tiburón as well. In desert scrub habitats, Seri individuals have seen it eat rodents and lizards. Despite its pugnacity, it is not dangerous to humans. The Seri who have been bitten while handling Gopher Snakes say that the bite barely hurts; nevertheless, young men urged me to kill one that I live-captured (and later released) at Tecomate. They collect the bones of this snake by picking up road kills or by actively seeking it out. The large vertebrae are used in necklaces, and are easily identified as Gopher Snake bones by virtue of their size (Jim Mead, pers. comm.).

Rhinocheilus lecontei
English common name: Long-nosed Snake
Spanish name: *coralillo falso*
Comcáac names: *coftj caacoj*

This slim, cream-colored snake with black and red blotches is likened to a large coralsnake by the Comcáac, as its name (*coftj,* "coralsnake," plus *caacoj,* "large") implies. It is known to frequent the same desert scrub habitats as Common

Kingsnakes, eating insects, toads, and small birds. María Luisa Astorga once killed one in dense mesquite scrub; it had been hidden by a tree. This nocturnal snake is widely but sporadically distributed, and is seldom seen within walking distance of the coastal villages. Craig Ivanyi and I have encountered it in ancient forests of ironwood and mesquite, and in saltbush desert scrub near abandoned fields between Calle 12 and Bahía Kino.

Salvadora hexalepis
English common name: Western Patch-nosed Snake
Spanish names: *alicante matorralera, culebra chata*
Comcáac names: *cocaznáacöl, maxáa*

This slender, striped snake is known by the Comcáac for its digging under trees and in sand, but they offer different names for it. They know it to be a fast-moving predator of dune and desert scrub habitats, where it eats mice, grasshoppers, and ants. Malkin (1962) learned from his Seri consultants that it is active mainly in the hot season, exhibits sexual dimorphism, and was once eaten by the Comcáac. It occurs on Isla Tiburón as well as on the mainland. The Comcáac know that it is not poisonous.

Trimorphodon biscutatus
English common name: Lyresnake
Spanish name: *víbora sorda*
Comcáac name: *(ziix) haas ano cocázni*

Known as a snake that prefers mesquite-covered rocky areas, the Lyresnake is called *haas ano cocázni*, "lives among mesquite like a rattlesnake." Ramón López Flores killed one at a cave named Hast Iinj near Desemboque, where it was preying on bats; herpetologists have also observed them climbing into bat roosts. The Comcáac claim that it preys upon land snails as well. Prey are killed by a venom from its teeth, according to what Seri consultants told Malkin (1962), but are also subdued by constriction. They can be active day or night, and are not particularly dangerous to humans.

ELAPIDAE: Coralsnakes

Micruroides euryxanthus
English common name: Western/Sonoran Coralsnake
Spanish name: *coralillo*
Comcáac name: *coftj*

Although false coralsnakes may be mistaken for it, this snake is always called *coftj* whenever it is seen. The Comcáac realize that this fossorial creature spends most of its time underground, in sandy or gravelly arroyos, or under the brushy canopies of desert scrub on sandy plains. It is known to hibernate during the winter, although Malkin (1962) was told that it had no seasonal preference. The Comcáac claim that it eats moths, larvae, toads, and lizards; herpetologists, however, have suggested that coralsnakes are selectively prey-directed toward blindsnakes, an interaction the Comcáac are unlikely to witness.

While most of the Comcáac understand that coralsnakes are venomous, they are unafraid of being bitten and regularly handle *coftj* (Malkin 1962). Several Comcáac elders simply stated that Seri individuals have never been killed by these gentle snakes. When they are found dead, their beautifully banded skin is made into good-luck rings or hat bands (Malkin 1962). This use has persisted, as I learned in October 2001, when elderly herbalist Juana Herrera presented me with a choker or hat band that she had fashioned from a coralsnake skin. She told me that this gift would bring me good luck.

Pelamis platurus
English common name: Yellow-bellied Seasnake
Spanish names: *alicante del mar, anguilla*
Comcáac name: *xepe ano cocázni*

This venomous serpent is sometimes confused with more frequently seen moray eels (probably *Gymnothorax castaneus*), which are called *xepe ano coimaj*, "rattleless serpent of the sea," by the Comcáac. However, a few elders know this seasnake, *xepe ano cocázni*, "rattlesnake of the sea," as the only aquatic serpent in the Sea of Cortés region. It is also the most widespread seasnake in the world (Minton 1989). Although this serpent's coloration is variously described by Seri elders, they all agree that it has tiny, needlelike teeth. Biologists consider the Yellow-bellied Seasnake to be extremely poisonous, but the Comcáac do not generally regard it to be dangerous unless it becomes entangled in nets. This lack of fear is all the more remarkable when placed in the global perspective that seasnakes probably kill more fishermen than does any other aquatic reptile.

Only one documented sighting of seasnakes is listed in the Sea of Cortés biodiversity database maintained in Guaymas, but biological historian Don Kucera (pers. comm.) has collected numerous other reports of their presence in the region. They have not been recently encountered by Comcáac fishermen in the Canal del Infiernillo, but are known from their visits to the shores of Baja California (Findley et al. 2001). At Punta Estrella near San Felipe, B.C., several Seri helped treat a Yellow-bellied Seasnake bite suffered by their relative Alfonso Méndez while he was pulling a net and line back into a panga. An entangled seasnake 30 centimeters long bit him on the foot. He was in extreme pain for three days as the wound continued to bleed, despite treatment with herbs such as *guaco* (*Wislizenia refracta*). He was then taken to a San Felipe clinic and ultimately to a hospital in Mexicali until his condition was stabilized. Two decades later, Alfonso told me that he still suffers daily from localized pain in his foot, and his nerves may be permanently impaired.

Crotalus angelensis and *C. estebanensis*
English common names: Isla Ángel de la Guarda Rattlesnake and Isla San Esteban
  Rattlesnake, respectively
Spanish names: *víbora de cascabel, víbora de las islas*
Comcáac name: *cocázni (sin yeesc)*

Although not given specific names in Cmique Iitom, these two distinct island populations of *cocázjc* (plural) are known for their peculiar rattle morphologies and unusually high densities (Grismer 1999). The Comcáac are aware that island rattlers can be functionally rattleless due to deformation in rattle segments, which allows them to slip off. Roberto Camposano used to kill dozens of these rattlers while he was fishing on the islands, bringing their carcasses home to process their bones for necklaces.

Crotalus atrox, C. molossus, and *C. scutulatus*
English common names: Western Diamondback Rattlesnake, Black-tailed
  Rattlesnake, and Mohave Rattlesnake, respectively
Spanish name: *víbora de cascabel*
Comcáac name: *cocázni*

The Western Diamondback is extremely variable in size, color, and markings within Comcáac homelands, as are the Mohave Rattlesnake and, to a lesser extent, the Black-tailed Rattlesnake (Campbell and Lamar 1989; Degenhardt, Painter, and Price 1996). Consequently, little effort is made to differentiate "kinds" or "types" of *cocázjc*, especially in moments of danger. *Cocázni* (pl. *cocázjc*) is used generically for all rattlers, but more specifically for Western Diamondbacks and Mohave Rattlers. The Comcáac do, however, comment on distinctive color morphs, such as a bright green Mohave we saw in dense brush near Campo El Caracol on Isla Tiburón in October 2001.

These rattlesnakes are ubiquitous in the desert scrub of the coastal plains but venture into rocky upland habitats as well. The Comcáac claim that they emerge from hibernation in mid-March, when they aggregate in caves in groups of twenty to twenty-five individuals (Malkin 1962). At Ensenada del Perro, Craig Ivanyi identified a Western Diamondback sharing a burrow with a Desert Tortoise, which Guadalupe López observed under an ironwood tree. According to Comcáac elders and children interviewed by Rosenberg (1997), rattlers will prey on small jackrabbits, cottontails, packrats, squirrels, lizards, and a variety of invertebrates.

Rattlesnakes are opportunistically killed to remove them from areas around family dwellings, but they are also sought out and harvested for their meat, oil, bones, and rattles. As María del Carmen Díaz explained to me, she and others kill rattlesnakes with pointed sticks to get their oil for medicinal purposes. The oil is extracted by cooking the snake meat like pork cracklings, then skimming the oil off the boiling water with a spoon. The oil from this and from *ctaamjij*, "Sidewinder," is used to treat earaches and skin wounds as well as to soothe a baby's mouth when it is sore from nursing. Felger and Moser (1985, 112) also reported rattlesnake fat being put on open festering sores and snakeskins being wrapped around a person's arms or chest to treat an area of internal pain.

Rattlesnake meat is still eaten as food by a few contemporary Seri, but more families use it medicinally. After it is roasted, the meat may be finely ground along with the bones, then added to the food of a victim diagnosed with tuberculosis (blood-vomiting disease). Rattlesnake meat is also cooked, dried, and pulverized, then eaten to cure ulcers, to treat cancers and arthritis, and to treat open wounds or sores that will not heal (Rosenberg 1997; Reina-Guerrero and Morales 1997). It is best not to provoke a rattler when killing it, Seri individuals claim, for that will contaminate the meat. The head and tail must be cut off before cooking.

The meat and the oil are now sold by a few Seri women to non-Indians who practice traditional medicine. These women claim that hearing loss can be treated by dusting impaired ears with the powder obtained from grinding down the rattles.

The rattles, bones, and skin are also used in necklaces and other personal ornamentation. The shed skin is called *cocázni ano yapox*, with *yapox* etymologically related to *coopox*, "to shed, to leave behind." Several songs about *cocázni* have been recorded by Comcáac singers.

Crotalus cerastes
English common name: Sidewinder
Spanish name: *cuernitos*
Comcáac name: (*cocázni*) *ctaamjij*

This peculiar lowland-dwelling rattler could be identified by photo as *ctaamjij* by 65 percent of Seri adults over forty years of age but only 50 percent of those under forty, the rest of whom offered only that it was a kind of *cocázni*. Those who distinguish *ctaamjij* from other rattlers mention its horns, its small size, and its movements as being distinctive. Although stereotyped as a dweller only of sandy washes and dunes, Sidewinders are also found on rocky beaches and adjacent bedrock benches in Comcáac homelands. Along rocky beaches and on the edge of tidepools, the Comcáac have seen Sidewinders eat intertidal isopods. They are considered to be nocturnal, hot-season-active creatures that produce live young (Malkin 1962). The Comcáac are also aware that their age can be established by counting the number of rattle segments on the tail (Malkin 1962). They are known to hibernate and aggregate in particular areas, with "families" of twenty to twenty-five individuals found within close proximity to one another.

Like Western Diamondback and Mohave Rattlers, Sidewinders are harvested for their meat, oil, vertebrae, and rattles. Whereas other rattlers are found on several of the midriff islands (in addition to the mainland), Sidewinders are found only on Isla Tiburón.

Crotalus tigris
English common name: Tiger Rattlesnake
Spanish names: *cascabel del tigre, víbora de cascabel*
Comcáac names: *cocázni cahtxíma, cocázni quiham*

Very few contemporary Seri individuals specifically distinguish this rattle-snake from others. *Cahtxíma* means "rich in color," an apt name for this snake. The adjective *quiham*, "immature," may have been specific to a particular snake we showed Alfredo López, or he may have called it that because of the small head size characteristic of Tiger Rattlesnakes. When Richard Nelson and I showed Alfredo a Tiger that had been killed by a Sonoran Coachwhip at Ensenada del Perro on Isla Tiburón, he did not call it anything more than *cocázni* until we offered the modifiers *cahtxíma* and *quiham*, which he concurred with. It was clear that he was familiar with its head shape and other distinguishing characteristics.

CROCODYLIDAE: Crocodiles and Relatives

Crocodylus acutus
English common name: River Crocodile
Spanish names: *caimán, cocodrilo*
Comcáac names: *xepe ano heepni, xepe ano paaza*

These Comcáac names are derived from the term for ocean or sea, *xepe*, and the names of two similarly shaped kinds of large land lizards, the spiny-tailed iguana and the Gila Monster. Although few Sonorans today know that River Crocodiles once inhabited their coastal waters, elderly Comcáac can recall several sightings of them within the last century. Two sightings of vagrants occurred in the 1900s, one of a large adult washed up on the southeast shores of Isla Tiburón, the second of a small juvenile seen in open water between Punta Mala on Tiburón and Punta Sargento. Seri individuals have recently visited zoos that keep crocodiles and verify that the morphology and size of these animals are consistent with what their forefathers described in animals that once were found as far north as Punta Sargento, near Empalme and Kino, the northernmost mangrove lagoon on the Sonoran coast. Alfredo López has also heard that the Comcáac infrequently sighted River Crocodiles on Islas San Esteban and San Lorenzo. This oral history is consistent with Felger's (1998) memoir of "lost crocodiles" at the southern edge of the Sonoran Desert coast, and with his personal communication from Edward Spicer, who sighted one between the Río Mayo and the Río Yaqui during the 1930s. Prior to the damming of all major rivers flowing into the Sea of Cortés, coastal mangrove lagoons no doubt formed much more continuous habitat suitable for crocodile nesting and feeding. This era probably ended in Sonora by the 1940s.

The elderly Seri who remember oral histories regarding crocodiles claim that these creatures were so big and fierce that they could kill sea lions and eat them. They were not eaten by the Comcáac, but the historic use of their hides is not implausible. Crocodiles still occur on the Nayarit coast and could be reintroduced to Sonora if translocation were linked to mangrove habitat restoration and constraints on hunting.

OTHER SPECIES

Becky Moser and her colleagues have recorded names for several other reptiles that remain to be identified with any confidence. One unidentified lizard is known as *hant caaxat*, meaning something akin to "being immersed in a dust plume from the earth." When it runs, the Comcáac say, a smokelike dust seems to come out of it and no one can get close to it. My first guess is that this might be the sand-loving Lesser Earless Lizard, *Holbrookia maculata*, and yet my Seri consultants could not identify it from preserved specimens. Then there is *hant ano coocöz*, a mythic serpent that causes earthquakes, just as the O'odham's *ñeibig* causes earthquakes, seepholes, and ground subsidence. The name *ziix coimaj hant cöquiih*, or serpent land-what-is-on, is probably a higher life-form category for non-venomous snakes rather than the name of a particular folk taxon. A sea turtle name, *stac casíi*, meaning "one that has the odor of pumice stone," likewise remains uncorrelated with any particular sea turtle.

reptile specimen records
from the sonoran coast
and nearby islands

Species	Sonoran Coast	Isla Tiburón	Other Islands
TURTLES AND TORTOISES			
TESTUDINIDAE			
Gopherus agassizii	Punta Hona: Lowe 1954, UAZ 36484	Lowe 1954, UAZ 36482, 36493; Rodríguez-Ruiz 1979	Dátil: Felger 1966; Grismer 1994a; questionable in Grismer 1999
KINOSTERNIDAE			
Kinosternon flavescens	Desemboque N (on beach after flood of Río Asunción): Brown 1979, UAZ 43037[1]		
CHELONIIDAE			
Caretta caretta	Bahía Kino: Findley et al. 2001		Ángel de la Guarda: Grismer 1994a
Chelonia mydas	Punta Chueca, Campo Egipto, Desemboque: Nabhan notes, photos 1998–1999		Ángel de la Guarda, San Esteban, San Lorenzo Norte y Sur, San Pedro Mártir: Grismer 1994a
Eretmochelys imbricata	Bahía Kino: Findley et al. 2001; Govan notes 1999		

Species	Sonoran Coast	Isla Tiburón	Other Islands
Lepidochelys olivacea	Puerto Libertad: Montgomery and Turk-Boyer in Findley et al. 2001	Grismer 1994a	San Pedro Mártir: Grismer 1994a
DERMOCHELYIDAE			
Dermochelys coriacea	Photos in this book	Grismer 1994a	

LIZARDS

CROTAPHYTIDAE

Crotaphytus dickersonae		Ensenada Blanca W side: Lowe 1963, UAZ 9625–26; Rodríguez-Ruiz 1979; Grismer 1994a, 1999	
C. nebrius	Desemboque 6.5 mi. NW: Lowe 1954, UAZ 694; Punta Cirio: Robinson 1967, UAZ 20144		
Gambelia wislizenii	Bahía Kino 16 mi. N: Lowe et al. 1968, UAZ 24220; Desemboque 4.7 mi. S: Bezy 1966, UAZ 15303	El Sauzal arroyo SW side: Felger 1958, UAZ 158990; Palo Fierro: Felger 1964, UAZ 22346; Grismer 1994a, 1999; Rodríguez-Ruiz 1979	
IGUANIDAE			
Ctenosaura conspicuosa			San Esteban: Lowe 1965, UAZ 35421; Cholludo: Felger 1958, UAZ 31114–5; San Esteban, Cholludo: Grismer 1999
C. nolascensis			San Pedro Nolasco E side: Robinson et al. 1974, UAZ 38588–93; Grismer 1999
Dipsosaurus dorsalis	Desemboque: Bezy 1966, UAZ 15187; Bahía Kino 1 mi.		

Species	Sonoran Coast	Isla Tiburón	Other Islands
	N: Cross 1969, UAZ 30864–65		
Sauromalus hispidus			Alcatraz: Lowe 1969, UAZ 32066 (w/ 32065, a hybrid); Ángel de la Guarda, San Lorenzo Norte y Sur: Grismer 1994a, 1999
S. obesus	Pozo Coyote 3 mi. N: Lowe 1951, UAZ 2232; Tastiota 1 mi. E: Wright 1963, UAZ 10511; Punta Cirio: Lowe 1961, UAZ 4637–38	Ensenada Blanco W side (tail only): Lowe 1963, UAZ 9707; Grismer 1994a, 1999; Rodríguez-Ruiz 1979	Alcatraz: Lowe 1954, UAZ 22456 (possibly a hybrid)
S. varius and *S. varius* × *S. obesus* × *S. hispidus*			San Esteban: Bezy 1968, UAZ 31149–50; San Esteban: Felger and Russell 1965, UAZ 35440; Alcatraz (hybrid): Keasey 1968, UAZ 29776; Grismer 1994a, 1999

PHRYNOSOMATIDAE

Species	Sonoran Coast	Isla Tiburón	Other Islands
Callisaurus draconoides	Desemboque: Felger 1963, UAZ 3811315; Tastiota: Lowe 1963, UAZ 3813538; Las Cuevitas 10 mi. N Desemboque: Bezy 1966, UAZ 15395407	Ensenada del Perro: Bezy 1968, UAZ 2374653; Coralito: Lowe 1954, UAZ 339; Tecomate: Lowe 1964, UAZ 2210506; Grismer 1994a, 1999; Rodríguez-Ruiz 1979	Patos: Grismer 1994a, 1999; Ángel de la Guarda: Grismer 1999
Holbrookia maculata	Vaughn 2002		
Phrynosoma solare		Grismer 1994a, 1999; Rodríguez-Ruiz 1979	
Sceloporus clarkii	Guaymas 4 km N: Lowe 2401; San Carlos 5 km N:	Grismer 1994a, 1999; Rodríguez-Ruiz 1979; Ensenada del Perro:	San Pedro Nolasco: Sherbrooke and Soulé 14273–75; Grismer

Species	Sonoran Coast	Isla Tiburón	Other Islands
	Robinson 24826	Ivanyi and Nabhan photos 1999	1999
S. magister	Desemboque: Bezy 1966, UAZ 15188; Campo Dolar/ Punta Sargento: Bezy 1966, UAZ 15312; Bahía San Carlos 2 mi. E: Lowe 1958, UAZ 2798	Palo Fierro: Felger 1965, UAZ 10644; Santa Rosa trail to SE Sierra Kunkaak: Felger 1963, UAZ 22463; Grismer 1994a, 1999; Rodríguez-Ruiz 1979	
Urosaurus graciosus	Puerto Libertad 1–4 mi. S: Robinson 1968, UAZ 24824; Desemboque: Felger 1963, UAZ 38110		
U. ornatus	Sierra Seri 18 mi. S Desemboque: Bezy 1966, UAZ 15326– 27; Bahía San Carlos: Harris 1960, UAZ 3872–73	Ensenada de la Cruz: Lowe 1958, UAZ 22599; upper central valley: Felger 1965, UAZ 22600; Grismer 1994a, 1999; Rodríguez-Ruiz 1979	
Uta stansburiana	Bahía Kino: Van Rossem 1947, UAZ 701314; Punta Sargento: Bezy 1966, UAZ 15322	El Sauzal water hole: Lowe 1954, UAZ 22971; Grismer 1994a,1999; Rodríguez-Ruiz 1979	San Esteban: Asplund and Shallenberger 1967, UAZ 19050; Alcatraz: Felger 1958, UAZ 2270516; Patos: Lowe 1963, UAZ 2271730; Dátil: Lowe 1958, UAZ 2293438; Ángel de la Guarda, Dátil, San Lorenzo Norte y Sur: Grismer 1999
EUBLEPHARIDAE Coleonyx variegatus	Desemboque 2 mi. S: Bezy 1968, UAZ 24822; Hermosillo 23 mi. W: Ferguson 1962, UAZ 11249	El Sauzal 2 mi. S: Felger 1965, UAZ 22058; Grismer 1994a; Rodríguez- Ruiz 1979	Ángel de la Guarda: Grismer 1994a, 1999

Species	Sonoran Coast	Isla Tiburón	Other Islands
GEKKONIDAE			
Phyllodactylus xanti		Ensenada Blanca W side: Felger 1965, UAZ 9703; Tecomate 12 mi. S: Lowe 1963, UAZ 2207374; Grismer 1994a, 1999; Rodríguez-Ruiz 1979	San Esteban: Asplund and Shallenberger 1967, UAZ 18092; Alcatraz, Cholludo, San Esteban: Grismer 1999; Ángel de la Guarda: Grismer 1994a
TEIIDAE			
Cnemidophorus tigris	Campo Víboras: Lowe 1954, UAZ 6552; Bahía San Carlos 3 km N: Robinson 1968, UAZ 10370	Ensenada Blanca W side: Lowe 1963, UAZ 25475; Grismer 1994a, 1999; Rodríguez-Ruiz 1979	San Esteban: Grismer 1994a; SE side: Lowe 1963, UAZ 9576–78; San Pedro Mártir: Grismer 1994a; Gilligan 1975, UAZ 40751–4; Ángel de la Guarda, San Lorenzo Norte y Sur, San Pedro Mártir: Grismer 1994a, 1999
HELODERMATIDAE			
Heloderma suspectum	Desemboque: Felger and Moser 1963, UAZ 36661; Caborca 35 mi. S at Cerro Jojoba: Felger 1963, UAZ 10459		
SNAKES			
BOIDAE			
Boa constrictor	Lowe 1961; Bahía San Carlos, El Palmar: Nabhan and Monti obs., 1999		
Charina trivirgata		Grismer 1994a, 1999; Rodríguez-Ruiz 1979	Ángel de la Guarda: Grismer 1999
COLUBRIDAE			
Arizona elegans	Bahía Kino 6 mi. E: Weiwandt 1969, UAZ 31423; Puerto Lobos 35 mi. SW		

Species	Sonoran Coast	Isla Tiburón	Other Islands
	Caborca: Lowe 1960, UAZ 24079; Hermosillo 24 mi. W: Wright 1963, UAZ 9642; Calle 12: Ivanyi and Nabhan photos, 1999		
Chilomeniscus stramineus	Tastiota N side: Holm 1995, UAZ 50630; Puerto Libertad 12 km N: Holm 1994, UAZ 50636	Grismer 1994a, 1999; Rodríguez-Ruiz 1979; El Sauzal arroyo S side: Felger, Moser, and Moser 1964, UAZ 23195	
Chionactis occipitalis	Puerto Lobos 1.1 km NNW Cerro Prieto: Holm and Rose 1995, UAZ 50629; Desemboque: Moser 1968, UAZ 29283; Punta Chueca: Nabhan and Ivanyi notes, 1998		
Hypsiglena torquata	Bahía Kino: Craig 1962, UAZ 39596; Kino to Calle 12: Ivanyi and Nabhan photos, 1999	Grismer 1994a, 1999; M. Vaughn, B. Wirt, G. Nabhan photo and notes, 2000	Ángel de la Guarda, San Esteban, San Lorenzo Sur: Grismer 1994a, 1999
Masticophis bilineatus	Guaymas 20 mi. E: Lowe 1956, UAZ 20805; Bahía Kino: Thompson 1972, UAZ 35058	Ensenada del Perro: Nabhan and Nelson photos, notes, 1998; Grismer 1994a, 1999; Rodríguez-Ruiz 1979	
M. flagellum	Desemboque: Felger 1964, UAZ 25647; Punta Chueca 1.5 mi. S: Lowe 1969, UAZ 31997	Palo Fierro E side: Felger 1964, UAZ 23220; Ensenada de la Cruz: Felger 1968, UAZ 23221; Grismer 1994a, 1999; Rodríguez-Ruiz 1979	Dátil: Grismer 1994a, 1999

Species	Sonoran Coast	Isla Tiburón	Other Islands
M. slevini			San Esteban: Felger 1968, UAZ 23038; Grismer 1994a, 1999
Phyllorhynchus decurtatus	Kino to Calle 12: Ivanyi and Nabhan photos, 1999		Ángel de la Guarda: Grismer 1994a
Pituophis melanoleucus	Tastiota 3.5 mi. E: Lowe 1962, UAZ 26054; Rancho Noche Bueno 4.6 mi. E: Felger 1963, UAZ 26078		
Rhinocheilus lecontei	Bahía San Carlos 2 mi. E: Lowe 1958, UAZ 4557; Calle 12 to Kino, Ivanyi and Nabhan photos, 1999		
Salvadora hexalepis		Grismer 1994a, 1999; Rodríguez-Ruiz 1979	
Trimorphodon biscutatus		Ensenada Blanca (Vaporeta) SW side: Felger 1964, UAZ 23291; Grismer 1994a, 1999; Rodríguez-Ruiz 1979	
ELAPIDAE Micruroides euryxanthus		Grismer 1994a, 1999; M. Vaughn and Nabhan notes 2001	
HYDROPHIIDAE Pelamis platurus	Guaymas/Miramar 1 mi. inland!: Welhorst 1962, UAZ 39726	Ensenada del Perro: Nabhan photos, notes, 1997	San Pedro: Peterson 1982, UAZ 44780; San Esteban: Nabhan photos, notes, 1997
VIPERIDAE Crotalus atrox	Puerto Libertad 3 mi. S: Lowe 1954, UAZ 27365; Desemboque: Bezy 1966, UAZ 32345	Ensenada Blanca (Vaporeta) SW side: Felger 1965, UAZ 23289; Grismer 1994a, 1999;	Dátil: Lowe 1954, UAZ 36631; Grismer 1994a, 1999; San Pedro Mártir: Keasey 1968, UAZ 28397; Grismer

Species	Sonoran Coast	Isla Tiburón	Other Islands
		Rodríguez-Ruiz 1979; Nabhan and Ivanyi photos, notes, 1998	1994a, 1999
C. cerastes	Bahía Kino 0.5 km N: Felger 1964, UAZ 27526; Desemboque 5.5 mi. N; Lowe, Robinson, and Soulé 1968, UAZ 27527	Lowe 1964, UAZ 23293, 28268–69; Grismer 1994a, 1999; Rodríguez-Ruiz 1979	
C. estebanensis			San Esteban: Grismer 1994a, 1999
C. molossus	Tastiota 10 mi. S: Bezy 1962, UAZ 13544	El Sauzal water hole: Russell 1963, UAZ 13660; Grismer 1994a, 1999; Rodríguez-Ruiz 1979	
C. ruber			Ángel de la Guarda: Grismer 1994a, 1999
C. scutulatus	Puerto Libertad 27.4 mi. S: Lowe 1954, UAZ 27779	M. Vaughn, B. Wirt, and G. Nabhan, photo and notes 2001	
C. tigris	Guaymas 17 mi. N: Soulé 1965, UAZ 27843	SE side, 1 mi. SE of (El Sauzal?) water hole: Felger 1963, UAZ 38206; Ensenada del Perro: Nabhan and Nelson photos, notes, 1998; Grismer 1994a, 1999; Rodríguez-Ruiz 1979	
CROCODYLIDAE			
Crocodylus acutus	Felger 1998	Nabhan oral history notes, 1997	

Sources: Grismer 1994a, 1999; Rodríguez-Ruiz 1979; Findley et al. 2001.

[1] The specimen numbers and collectors cited refer to preserved specimens registered in the University of Arizona herpetological collection prior to summer 1998. Other collections-based inventories that were used to compile this appendix include those of Arizona State University, the Arizona-Sonora Desert Museum, and the San Diego Natural History Museum.

literature cited

Astorga, A., G. P. Nabhan, and J. Miller. 2001. *Efraín of the Sonoran Desert: A Lizard's Life among the Seri Indians*. El Paso: Cinco Puntos Press.

Austin, C. C. 1999. Lizards took express train to Polynesia. *Nature* 397 (6715): 113–114.

Bahr, D. M., J. Gregorio, D. López, and A. Alvarez. 1979. *Piman Shamanism and Staying Sickness (Ka:cim Mumkidag)*. Tucson: University of Arizona Press.

Bahre, C. 1983. Human impacts: The midriff islands. In *Island Biogeography of the Sea of Cortez*, ed. T. Case and M. Cody, 290–305. Berkeley: University of California Press.

Bahre, C. J., L. Bourillón, and J. Torre. 2000. The Seri and commercial totoaba fishing (1930–1965). *Journal of the Southwest* 42 (3): 559–574.

Beals, R. 1945. Modern serpent beliefs in Mexico. In *The Contemporary Culture of the Cahita Indians*. Bureau of American Ethnology, Bulletin 142. Washington, D.C.: U.S. Government Printing Office.

Berkes, F. 1999. *Sacred Ecology: Traditional Ecological Knowledge and Resource Management*. Philadelphia: Taylor & Francis.

Berlin, B. 1992. *Ethnobiological Classification: Principles of Categorization of Plants and Animals in Traditional Societies*. Princeton: Princeton University Press.

Bowen, T. 2000. *Unknown Island: Seri Indians, Europeans, and San Esteban Island*. Albuquerque: University of New Mexico Press.

Brown, D., ed. 1982. Biotic communities of the American Southwest—U.S. and Mexico. *Desert Plants* 4 (1–4): 1–342.

Buege, D. J. 1996. The ecologically noble savage revisited. *Environmental Ethics* 18: 71–88.

Burckhalter, D. 1999. *Among Turtle Hunters and Basket Makers—Adventures with the Seri Indians.* Tucson: Treasure Chest Books.

Bury, R. B., D. J. Germano, M. C. Melendez, and C. R. Schwalbe. 2002. Distribution, ecology, and conservation of the Desert Tortoise in Mexico. In *The Sonoran Desert Tortoise,* ed. T. R. Van Devender. Tucson: University of Arizona Press.

Bye, R. A., Jr., and M. L. Zigmond. 1976. Review of *Principles of Tzetzal Plant Classification,* by B. Berlin, D. Breedlove, and P. Raven. *Human Ecology* 4 (3): 171–175.

Calvino, I. 1983. *Mr. Palomar.* Translated by W. Weaver. New York: Harcourt Brace Jovanovich.

Campbell, J. A., and W. L. Lamar. 1989. *The Venomous Reptiles of Latin America.* Ithaca: Comstock Publishing Associates.

Case, T. J., and M. L. Cody. 1987. Island biogeographies: Tests on islands in the Sea of Cortés. *American Scientist* 75: 402–411.

————, eds. 1983. *Island Biogeography in the Sea of Cortez.* Berkeley: University of California Press.

Centro Ecológico de Sonora. 1994. Plan de manejo: Reserva de Biosfera Islas Tiburón y San Esteban. Draft proposal submitted to SEDESOL, Mexico City.

Clifford, J. 1981. On ethnographic surrealism. *Comparative Studies in Society and History* 23: 539–564.

Cliffton, K. D. 1990. Leatherback turtle slaughter in Mexico. *Sonoran Herpetologist* 3 (5): 44–46.

Cliffton, K. D., O. Cornejo, and R. S. Felger. 1982. Sea turtles of the Pacific coast of Mexico. In *Biology and Conservation of Sea Turtles,* ed. K. A. Bjorndal, 199–209. Washington, D.C.: Smithsonian Institution Press.

Coolidge, D. 1981. *Texas Cowboys.* Tucson: University of Arizona Press.

Coolidge, D., and M. R. Coolidge. 1971. *The Last of the Seris.* Glorieta, N.M.: Rio Grande Press.

Cox, P. A. Indigenous peoples and conservation. In *Biodiversity and Human Health,* ed. F. Grifo and J. Rosenthal, 207–220. Washington, D.C.: Island Press.

Crystal, D. 2000. *Language Death.* Oxford: Oxford University Press.

Cudney-Bueno, R., and P. Turk-Boyer. 1998. *Pescando entre mareas del alto Golfo de California.* Puerto Peñasco, Sonora: CEDO.

Darwin, C. 1997. *Voyage of the Beagle.* New York: Penguin Books.

Degenhardt, E. G., C. W. Painter, and A. H. Price. 1996. *Amphibians and Reptiles of New Mexico.* Albuquerque: University of New Mexico Press.

DiPeso, C. C., and D. S. Matson. 1965. The Seri Indians in 1692 as described by Adamo Gilg, S.J. *Arizona and the West* 7 (1): 33–56.

Eckert, S. 1999. *Habitats and Migratory Pathways of the Pacific Leatherback Sea Turtle.* Hubbs Sea World Institute Technical Report 99-290.

Edwards, J. 1995. *Multi-lingualism.* New York: Penguin Books.

Espinosa-Reyna, A. 1997. *La historia en el rostro.* Hermosillo: Publicaciones del Gobierno del Estado de Sonora, Secretaria de Educación y Cultura.

Felger, R. S. 1965. *Xantusia vigilis* and its habitat in Sonora, Mexico. *Herpetologica* 21: 146–147.

———. 1990. The Seri Indians and their herpetofauna. *Sonoran Herpetologist* 3 (5): 41–44.

———. 1998. Lost crocodiles in Sonora. *Drylands Oasis* 1 (1): 2–7.

———. 1999. The flora of Cañón de Nacapule: A desert-bounded tropical canyon near Guaymas, Sonora, Mexico. *Proceedings of the San Diego Society of Natural History* 35: 1–42.

Felger, R. S., K. Cliffton, and P. J. Regal. 1978. Winter dormancy in sea turtles: Independent discovery and exploitation in the Gulf of California by two local cultures. *Science* 191: 283–285.

Felger, R. S., and M. B. Moser. 1974. Seri Indian pharmacopeia. *Economic Botany* 28 (4): 414–436.

———. 1975. Columnar cacti in Seri Indian culture. *The Kiva* 39 (3–4): 257–275.

———. 1985. *People of the Desert and Sea: Ethnobotany of the Seri Indians.* Tucson: University of Arizona Press.

———. 1987. Sea turtles in Seri Indian culture. *Environment Southwest* 519: 18–23.

Felger, R. S., M. B. Moser, and E. W. Moser. 1983. The Desert Tortoise in Seri Indian culture. *Proceedings of the Desert Tortoise Council, 28–30 May 1981, Riverside, Ca.,* 6: 113–120.

Ferreira, L. A. F., et al. 1992. Snakebite medicines. *Toxicon* 30 (10): 1211–1218.

Figueroa, A. L. 1999. *Programa de manejo, Islas del Golfo de California.* Guaymas: SEMARNAP/ ISLAS.

Findley, L., et al. 2001. Gulf of California Macrofauna Diversity Project Data Base. Conservation International/Mexico, Guaymas, Sonora. [On CD.]

Fisher, R. N. 1997. Dispersal and evolution of the Pacific Basin gekkonid lizards, *Gehyra oceanica,* and *Gehyra mutilata. Evolution* 51 (3): 903–921.

Fowler, G. 1997. *Mystic Healers and Medicine Shows.* Santa Fe, N.M.: Ancient City Press.

Fritts, T. H., M. L. Stinson, and R. Marquez M. 1982. Status of sea turtle nesting in southern Baja California, Mexico. *Bulletin of the Southern California Academy of Sciences* 81 (2): 51–60.

Gentry, H. S. 1938. Herbarium specimen notes from San Joaquín, Baja California. University of Arizona Herbarium, Tucson.

González-Romero, A., and S. Alvarez-Cardenas. 1989. Herpetofauna de la región del Pinacate, Sonora, México: un inventario. *Southwestern Naturalist* 34 (4): 519–526.

Green, R. 1981. Culturally based science: The potential for traditional people, science, and folklore. In *Folklore in the Twentieth Century*, ed. M. Newall, 204–212. London: Rowman & Littlefield.

Grenard, S. 1994. *Medical Herpetology*. Pottsville, Pa.: NPG Publishing.

Griffen, W. B. 1959. *Notes on Seri Indian Culture, Sonora, Mexico*. School of Inter-American Studies, Latin American Monograph Series 10. Gainesville: University of Florida Press.

Griffith, J. 1989. *Beliefs and Holy Places*. Tucson: University of Arizona Press.

Grismer, L. L. 1994a. Evolutionary and ecological biogeography of the herpetofauna of Baja California and the Sea of Cortés, Mexico. Ph.D. diss., Loma Linda University.

———. 1994b. Geographic origins for the reptiles on islands in the Gulf of California, Mexico. *Herpetological Natural History* 2 (2): 17–41.

———. 1999. Checklist of amphibians and reptiles on islands in the Gulf of California, Mexico. *Bulletin of the Southern California Academy of Sciences* 98 (1).

———. 2002. *Amphibians and Reptiles of Baja California, Including Its Pacific Islands and the Islands in the Sea of Cortés*. Berkeley: University of California Press.

Hames, R. 1991. Wildlife conservation in tribal societies. In *Biodiversity: Culture, Conservation, and Ecodevelopment*, ed. M. L. Oldfield and J. B. Alcorn, 172–199. Boulder, Colo.: Westview Press.

Hanski, I. 1999. An explosive laboratory. Review of *Island Biogeography: Ecology, Evolution and Conservation*, by Robert J. Whittaker. *Nature* 398 (6716): 115.

Harding, S. 1998. *Is Science Multicultural?* Bloomington: Indiana University Press.

Harmon, D. 1995. The status of the world's languages as reported in *Ethnologue*. *Southwest Journal of Linguistics* 14 (1–2):1–28.

———. 1996. The converging extinction crisis: Defining terms and understanding trends in the loss of biological and cultural diversity. Keynote address at the April 1, 1996, Colloquium "Losing Species, Languages, and Stories: Linking Cultural and Environmental Change in the Binational Southwest." Arizona-Sonora Desert Museum, Tucson.

Hills, J. 1973. An ecological interpretation of prehistoric Seri settlement patterns in Sonora, Mexico. Master's thesis, Arizona State University.

Hine, C. H. 2000. Five Seri spirit songs. *Journal of the Southwest* 42 (3): 589–611.

Hine, C. H., and J. Hills. 2000. Seri concepts of place. *Journal of the Southwest* 42 (3): 583–588.

Hinton, L., J. Nichols, and J. J. Ohala, eds. 1994. *Sound Symbolism*. New York: Cambridge University Press.

Hogan, L. 1995. Creations. In *Dwellings*. New York: W. W. Norton.

Hollingsworth, B. D., C. R. Mahralt, L. L. Grismer, B. H. Bantu, and C. K. Silber. 1997. The occurrence of *Sauromalus varius* on a satellite islet of Salsipuedes, Gulf of California, Mexico. *Herpetological Review* 28 (1): 26–28.

Houghton, P. J. 1994. Treatment of snakebite with plants. *Journal of Wilderness Medicine* 5: 451–452.

Houghton, P. J., and I. M. Osibogen. 1993. Flowering plants used against snakebite. *Journal of Ethnopharmacology* 39: 1–29.

Hunn, E. 1977. *Tzetzal Folk Taxonomy*. New York: Academic Press.

Johnson Gordon, D., S. F. Moreno Salazar, and R. López Estudillo. 1994. *Compendio fitoquímico de la medicina tradicional herbolaria de Sonora*. Hermosillo: University of Sonora.

Jones, K. B., and W. G. Whitford. 1989. Feeding behavior of free-roaming *Masticophis flagellum*: An efficient ambush predator. *Southwestern Naturalist* 34 (4): 460–467.

Kellert, S., and E. O. Wilson, eds. 1993. *The Biophilia Hypothesis*. Washington, D.C.: Island Press.

Kloppenburg, J. R. 1992. Science in agriculture. *Rural Sociology* 57 (1): 96–107.

Kroeber, A. 1931. *The Seri*. Southwest Museum Papers, 6. Los Angeles: Southwest Museum.

Linden, E. 1991. Lost tribes, lost knowledge. *Time*, Sept. 23.

Lowe, C. H., Jr. 1959. Contemporary biota of the Sonoran Desert: Problems. In *University of Arizona Arid Land Colloquium Proceedings, 1958*, 54–74.

Lowe, C. H., Jr., and K. S. Norris. 1955. New and revised subspecies of Isla de San Esteban, Gulf of California, Sonora, Mexico, with notes on other satellite islands of Isla Tiburón. Analysis of the herpetofauna of Baja California, Mexico. III. *Herpetologica* 11 (2): 89–96.

MacArthur, R. H., and E. O. Wilson. 1967. *The Theory of Island Biogeography*. Princeton: Princeton University Press.

Mace, G. M., and R. Lande. 1993. Assessing extinction threats: Toward a reevaluation of IUCN threatened species categories. *Conservation Biology* 4: 52–62.

Malkin, B. 1962. *Seri Ethnozoology*. Occasional Paper No. 7. Pocatello: Idaho State College Museum.

Mann, C. C., and M. L. Plummer. 1995. *Noah's Choice: The Future of Endangered Species*. New York: Alfred A. Knopf.

Marlett, S. A. 2000. Why the Seri language is important and interesting. *Journal of the Southwest* 42 (3): 611–635.

Marlett, S. A., and M. B. Moser. 1995. Presentación y análisis preliminar de 600 topónimos Seris. Paper presented at the conference "Topónimia: Los nombres de los pueblos del noroeste," Colegio de Sinaloa, Culiacán.

McGee, W. J. 1898. *The Seri Indians*. Seventeenth Annual Report of the Bureau of American Ethnology, Smithsonian Institution, Washington, D.C.

———. 1971. *The Seri Indians of Bahía Kino and Sonora, Mexico*. Glorieta, N.M.: Rio Grande Press.

McKeown, S. 1978. *Hawaiian Reptiles and Amphibians*. Honolulu: Oriental Publishing Company.

Mellado, A. n.d. *José Astorga, su historia*. Hermosillo, Sonora: INI.

Minton, S. A. 1989. The life and times of sea snakes. *Sonoran Herpetologist* 2 (1): 3–5.

Molina Molina, F. 1972. *Nombres indígenas de Sonora y su traducción al Español*. Hermosillo, Sonora: self-published.

Monti, L., ed. 1998. *Los Comcáac y su comida tradicional: Como prevenir el diabétis*. Tucson: Arizona-Sonora Desert Museum and Amazon Conservation Team.

Morris, R., and D. Morris. 1965. *Men and Snakes*. New York: McGraw-Hill.

Moser, E. 1963. Seri bands. *The Kiva* 28 (3): 14–27.

Moser, E., M. B. Moser, and S. A. Marlett. 2001. *Seri: Intercontinental Dictionary*. [On CD.] M. R. Key, general editor. A prepublication version appeared as (1996) Report 9: Survey of California and Other Indian Languages. Proceedings of the Hokan-Penutian Workshop, ed. Victor Golla, 191–232.

Mundkur, B. 1983. *The Cult of the Serpent*. Albany: State University of New York Press.

Nabhan, G. P. 1998a. Handing down ecological knowledge: Reviving an ancient sense of place among the Seri Indians. *Orion Afield* (fall): 28–31.

———. 1998b. Songs of the Seri. *Sierra*, Nov.-Dec., 2–9.

———. 2000a. Native American management and conservation of biodiversity in the Sonoran Desert bioregion: An ethnoecological perspective. In *Biodiversity and Native America*, ed. P. E. Minnis and W. J. Elisens, 29–43. Norman: University of Oklahoma Press.

———. 2000b. Interspecific relationships affecting endangered species recognized by O'odham and Comcáac cultures. *Ecological Applications* 10 (5): 1288–1295.

———. 2000c. Cultural dispersal of plants and reptiles to the Midriff Islands of the

Sea of Cortés: Integrating indigenous human dispersal agents into island biogeography. *Journal of the Southwest* 42 (3): 545–558.

———. 2001. Cultural perceptions of ecological interactions: An "endangered people's" contribution to the conservation of biological and linguistic diversity. In *Language, Knowledge, and the Environment: The Interdependence of Biological and Cultural Diversity*, ed. L. Maffi. Washington, D.C.: Smithsonian Institution Press.

———. 2002. When desert tortoises talk, Indians listen: Traditional ecological knowledge of Sonoran Desert cultures. In *The Sonoran Desert Tortoise*, ed. T. Van Devender. Tucson: University of Arizona Press.

Nabhan, G. P., and J. Carr, eds. 1994. *Ironwood: An Ecological and Cultural Keystone of the Sonoran Desert*. Conservation International Monographs in Conservation Biology 2. Chicago: University of Chicago Press.

Nabhan, G. P., H. Govan, S. A. Eckert, and J. A. Seminoff. 1999. Taller sobre tortugas marinas para el tribu indigena Seri. *Noticiero de tortugas marinas* 86: 14.

Nabhan, G. P., and M. Klett. 1995. *Desert Legends*. New York: Henry Holt.

Nabhan, G. P., and P. Mirocha. 1985. *Gathering the Desert*. Tucson: University of Arizona Press.

Nabhan, G. P., and L. Monti. 1997–1999. Field notes from Punta Chueca and Desemboque. Deposited in University of Arizona Library Special Collections, Tucson.

Nabhan, G. P., and S. St. Antoine. 1993. The loss of floral and faunal story: The extinction of experience. In *The Biophilia Hypothesis*, ed. S. R. Kellert and E. O. Wilson, 229–250. Washington, D.C.: Island Press.

Navarro, C. J. 1997. Cuentos del mar—nacimiento de caguamas. *CEDO News* 7 (4): 29–32.

Nelson, R. 1983. *Make Prayers to the Raven*. Chicago: University of Chicago Press.

Ohmagari, K., and F. Berkes. 1997. Transmission of indigenous knowledge and bush skills among the Western James Bay Cree women of subarctic Canada. *Human Ecology* 25 (2): 197–222.

Olson, S. L., and H. James. 1984. The role of Polynesians in the extinctions of the avifauna of the Hawaiian islands. In *Quaternary Extinctions: A Prehistoric Revolution*, ed. P. S. Martin and R. G. Klein, 768–780. Tucson: University of Arizona Press.

Parra-Rizo, J. H., and S. Virsano-Bellow. 1994. Por el camino culebrero: Etnobotánica y medicina de los indigenas Awa del Sábalo (Nariño). Bogotá: Ediciones Aba-Yala.

Petren, K., and T. J. Case. 1997. A phylogenetic analysis of body size evolution and biogeography of chuckwallas (*Sauromalus*) and other iguanids. *Evolution* 51 (1): 206–219.

Plotkin, M. 2000. *The Healer's Quest*. New York: Viking.

Quammen, D. 1996. *Song of the Dodo: Island Biogeography in an Age of Extinctions*. New York: Simon & Schuster.

Rapoport, E. H. 1982. *Areography: Geographical Strategies of Species*. New York: Pergamon Press.

Rea, A. M. 1997. *By the Desert's Green Edge*. Tucson: University of Arizona Press.

Reader, T. 1997. Recontextualizing culture: Some reflections on the material roots of language and culture. Sells, Ariz.: Tohono O'odham Community Action. Typescript.

Redford, K. H. 1989. The ecologically noble savage. *Orion* 9 (3): 24–29.

Reina-Guerrero, A.-L. 1994. *Medicina tradicional Seri*. Proceedings of the Eighth International Congress on Traditional Medicine and Folklore, Memorial University of Newfoundland, St. Johns, Canada.

Reina-Guerrero, A.-L., and D. Morales. 1997. *Medicina tradicional Seri*. Hermosillo, Sonora: Culturas Populares.

Reyes-Osorio, S. 1979. Aspectos biologicos de la tortuga del desierto (*Gopherus agassizii*) en la Isla Tiburón, Sonora. In *Memoria del IV simposio sobre el medio ambiente del Golfo de California*, 142–153. Instituto Nacional de Investigaciones Forestales, Publicación especial 17. Mexico City.

Reyes-Osorio, S., and R. B. Bury. 1982. Ecology and status of the desert tortoise (*Gopherus agassizii*) on Tiburón Island, Sonora. In *North American Tortoises: Conservation and Ecology*, ed. R. B. Bury, 39–49. U.S. Fish and Wildlife Service, Wildlife Research Report 12. Washington, D.C.

Richardson, M. 1972. *The Fascination of Reptiles*. New York: Hill & Wang.

Robinson, M. D. 1972. Chromosomes, protein polymorphisms, and systematics of insular chuckwalla lizards (genus *Sauromalus*) in the Gulf of California. Ph.D. diss., University of Arizona, Tucson.

Rodríguez-Ruiz, M. 1979. Estudios realizados en la estación experimental en zonas áridas de Isla Tiburón, Sonora. *Memoria del IIII Simposio Binacional sobre el Medio Ambiente del Golfo de California*, SARH Instituto Nacional de Investigaciones Forestales, no. 14, 69–72.

Romney, A. K., S. C. Weller, and W. H. Batchelder. 1986. Culture as consensus: A theory of culture and informant accuracy. *American Anthropologist* 88 (2): 313–338.

Rosenberg, J. 1997. Documenting and revitalizing traditional ecological knowledge: The curriculum development component of the Seri ethnozoology education project. Master's thesis, University of Arizona.

Rosenberg, J., P. Romero, G. P. Nabhan, and H. Lawler. 1997. *Animalitos del desierto y del mar*. Tucson: Arizona-Sonora Desert Museum.

Rzedowski, J. 1993. Diversidad y origenes de la flora phaenerogámica de México. *Acta Botanica Mexicana* 14: 3–21.

St. Antoine, S. 1994. Seri Indian arts and crafts. In *Ironwood: An Ecological and Cultural Keystone of the Sonoran Desert*, ed. G. P. Nabhan and J. Carr. Conservation International Monographs in Conservation Biology 2, 45–62. Chicago: University of Chicago Press.

Sarti, L. M., S. A. Eckert, N. T. Garcia, and A. R. Barragan. 1996. Decline of the world's largest nesting assemblage of leatherback turtles. *Marine Turtle Newsletter* 74: 2–5.

Sauer, C. 1978. *Seeds, Spades, Hearths, and Herds*. New York: Oxford University Press.

Sauer, J. 1969. Oceanic islands and biogeographic theory: A review. *Geographical Review* 59 (4): 583–589.

Seminoff, J. A., J. Alvarado, C. Delgado, J. L. López, and G. Hoeffer. 2002. First direct evidence of migration by an East Pacific green seaturtle from Michoacán, Mexico, to a feeding group on the Sonoran coast of the Gulf of California. *Southwestern Naturalist* 47 (2): 314–317.

Sevilla, G. 1998. Protecting rare lizard may help poor tribe, too. *Columbus Dispatch*, June 9.

Shepard, P. 1993. On animal friends. In *The Biophilia Hypothesis*, ed. S. Kellert and E. O. Wilson, 275–300. Washington, D.C.: Island Press.

Sherbrooke, W. C. 1981. *Horned Lizards: Unique Reptiles of Western North America*. Tucson: Southwest Parks and Monuments Association.

Sheridan, T. E. 1982. Seri bands in cross-cultural perspective. *The Kiva* 47 (4): 185–211.

———. 1987. Captive *Phrynosoma solare* raised without ants or hibernation. *Herpetological Review* 18: 11–15.

———. 1997. The Seri Indians. In *Paths of Life*, by N. Parezo and T. Sheridan. Tucson: University of Arizona Press.

———. 1999. *Empire of Sand: The Seri Indians and the Struggle for Spanish Sonora, 1645–1803*. Tucson: University of Arizona/Arizona State Museum.

Silber, C. H. 1985. Feeding habits, reproduction, and relocation of insular giant chuckwallas. Ph.D. diss., Colorado State University, Fort Collins.

Simberloff, D., J. A. Farr, J. Cox, and D. W. Mehlman. 1992. Movement corridors: Conservation bargains or poor investments? *Conservation Biology* 6 (4): 493–504.

Smith, W. N. 1951. Field notes on Seri culture. University of Arizona Library Special Collections, Tucson.

———. 1974. The Seri Indians and sea turtles. *Journal of Arizona History* 15 (2): 139–158.

Sobarzo, H. 1966. *Vocabulario sonorense*. Mexico City: Editorial Porrua.

Soulé, M. E. 1987. *Viable Populations for Conservation*. Cambridge: Cambridge University Press.

Soulé, M. E., and A. J. Sloan. 1966. Biogeography and distribution of the reptiles and amphibians on islands in the Gulf of California, Mexico. *Transactions of the San Diego Society of Natural History* 14 (11): 137–156.

Spotila, J. R., et al. 1996. Worldwide population decline of *Dermochelys coricea*: Are leatherbacks going extinct? *Chelonian Conservation and Biology* 2 (2): 209–222.

Steinbeck, J., and E. F. Ricketts. 1941. *The Sea of Cortez.* New York: Viking Press.

Thomas, C., and S. Scott. 1997. *All Stings Considered: First Aid and Medical Treatment of Hawaii's Marine Injuries.* Honolulu: University of Hawaii Press.

Thompson, J. N. 1996. Evolutionary ecology and the conservation of interaction diversity. *Trends in Ecology and Evolution* 11 (7): 300–303.

———. 1999. The evolution of species interactions. *Science* 284: 2116–2118.

Thomson, D. A., L. T. Findley, and A. N. Kerstitch. 1987. *Reef Fishes of the Sea of Cortez: The Rocky-Shore Fishes of the Gulf of California.* Tucson: University of Arizona Press.

Toledo, V. 1995. New paradigms for a new ethnobotany: Reflections on the case of Mexico. In *Ethnobotany: Evolution of a Discipline*, ed. R. E. Schultes and S. Von Reis, 75–93. Portland, Oreg.: Dioscorides Press.

Turner, N. R. 1996. *Islands of Home.* Portland, Oreg.: Ecotrust.

Van Devender, T. R., P. A. Holm, E. B. Wirt, B. E. Martin, C. R. Schwalbe, V. M. Dickinson, and S. L. Barrett. 2002. Grasses, mallows, desert vines, and more: Diets of the desert tortoise in Arizona and Sonora. In *The Sonoran Desert Tortoise*, ed. T. Van Devender. Tucson: University of Arizona Press.

Van Devender, T. R., C. H. Lowe, and H. E. Lawler. 1994. Factors influencing the distribution of the neotropical vine snake (*Oxybelis aeneus*) in Arizona and Sonora. *Herpetological Natural History* 2 (1): 25–42.

Vásquez-Dávila, M. A. 1997. El amash y el pistoque: Un ejemplo de la etnoecología de los Chontales de Tabasco, México. *Etnoecología* 3: 59–69.

Vaughn, M. 2002. Desert Tortoise survey by Comcáac para-ecologists, assisted by the Desert Tortoise Council, Isla Tiburón, September 2001. Unpublished report for NAU/CSG and Desert Tortoise Council, Flagstaff, Arizona.

Villapando, C. E. 1989. Los que viven en las montañas. *INAH Noroeste de México* 8: 1–104.

Whiting, A. 1951. A Seri diary, June 11 to 24, 1951. Field notes and correspondence to W. N. Smith. University of Arizona Library Special Collections, Tucson.

Williamson, M. 1989. The MacArthur and Wilson theory today: True but trivial. *Journal of Island Biogeography* 16: 3–4.

Wilson, E. O., and T. Taylor. 1967. An estimate of the potential evolutionary increase in species density in the Polynesian ant fauna. *Evolution* 21 (1): 1–10.

———. 1998. *Consilience: The Unity of Knowledge.* New York: Alfred A. Knopf.

Yetman, D. A., and A. Burquez. 1996. A tale of two species: Speculation on the introduction of *Pachycereus pringlei* in the Sierra Libre, Sonora, Mexico. *Desert Plants* 12 (1): 23–30.

Zent, S. 1999. The quandary of conserving ethnobiological knowledge: A Prioa example. In *Ethnoecology: Knowledge, Resources, and Rights*, ed. T. Gragson and B. Blount, 90–123. Athens: University of Georgia Press.

Zethelius, M., and M. J. Balick. 1982. Modern medicine and shamanistic ritual: A case of positive synergetic response in the treatment of a snakebite. *Journal of Ethnopharmacology* 5: 181–185.

Zitnow, J. D. 1990. A comparison of time Ojibway adolescents spend with parents/elders in the 1930s and 1980s. *American Indian and Alaskan Native Mental Health Research* 3 (3): 7–16.

Zuñiga, J. C. 1998. Se preserva la cultura Conca'ac. *El Imparcial*, Dec. 6, D-1.

Index

Page numbers in *italic* type refer to tables, figures, and plates. Page numbers in **boldface** type refer to the species accounts in Part 2.

Aboriginal introduction of species.
　See Cultural dispersal of species
Alcatraz. *See* Isla Alcatraz
Algae: with reptile-related names,
　165; on sea turtle carapaces, *65,*
　157, 166, 167, 245, 246; as sea
　turtle food, *166,* 240, 244
Alicante del mar. *See Pelamis platurus*
Alicante matorralera. *See Salvadora hexalepis*
Amazon Conservation Team, 195, 197,
　216
Ángel de la Guarda. *See* Isla Ángel
　de la Guarda
Anguilla. *See Pelamis platurus*
Arizona elegans (Glossy Snake), *74, 169,*
　172, **268,** 287
Arizona-Sonora Desert Museum, 84,
　85, *86,* 90, 150, 195, 216, 217, 222,
　235
Arts and crafts, 145; basketry, *124,*
　127, 203; carvings, 2, *7–9, 42, 79,*
　102, 108, 145–51, *147, 148, 149,*
　151, 159, plates 15,16; ceremonial

face painting, 146; connection with
　indigenous science, 3; reflecting
　Comcáac knowledge of reptiles,
　7–9, 149–50; sand drawing, *145,*
　plate 17; and tourism, 7, 145–51,
　189. *See also* Jewelry
Ashy Limberbush. *See Jatropha cinerea*
Astorga, Amalia, *iv,* 88, *127, 139, 160,*
　247, 270, Plates 9,10; on Birthing
　Place, *54, 96;* and Pozo Coyote, 27;
　and Efraín the Side-blotched Lizard,
　175–76, *262;* on sea serpent, *41–*
　42; and sea turtle incident, *151–52*
Astorga, Irma, *iv*
Astorga, José, *148*
Astorga, María Luisa, *iv, 274*
Astorga, Olga, *iv*
Astorga, Santiago, *iv*
Astorga, Victoria, *iv*
Athene cunicularia (Burrowing Owl), 228

Bahía Kino, *46,* 61, 89, 283, 284,
　286, 287, 288, 290; and commer-

cial fishing, 87–88; and market for
　turtles/tortoises, 187, 193; as turtle
　egg-laying site, 67, 68
Bahía San Carlos, 286, 287, 289
Baja California Rattlesnake. *See Crotalus*
　enyo
Bandless Sandsnake. *See Chilomeniscus*
　stramineus
Barefoot Banded Gecko. *See Coleonyx*
　gypsicolus
Barnet, Francisco "Chapo," *iv,* 44,
　162, 228
Barnet, Francisco "Pancho Largo," *iv*
Barnet, Ignacio "Nacho," *iv,* 9, *105,*
　124, 189, 217, 253, 263
Barnet, Martín Enrique, *iv*
Barnet, Miguel, *iv, 132*
Barnet, Ramona, *iv, 163*
Barnet, Raymundo, *iv*
Barrera, Narciso, *5*
Basketry, *124, 127, 203*
Baúla. *See Dermochelys coriacea*
Beals, Ralph, *58*

303

Beauty, reptiles associated with, 125–26, 128–29
Bejori. See *Sceloporus magister*
Bejuquillo. See *Oxybelis aeneus*
Berlin, Brent, 115–16
Biogeography, equilibrium theory of, 81–84
Biophilia, 122, 152, 174
Birds: endangered, 31–32; extinction of flightless species, 191; and place-names, 65
Birthing Place, 54, 95–96
Black Chuckwalla. See *Sauromalus hispidus*
Black Sea Turtle. See *Chelonia mydas*
Black-tailed Brush Lizard. See *Urosaurus nigricaudus*
Black-tailed Rattlesnake. See *Crotalus molossus*
Blanco, José Luis, iv, 1–2, 79, 108, 159, 267
Blanco, María Guadalupe, iv
Boa constrictor (Boa Constrictor), 47, 74, 104, 129, 169, 172, **266–67**, 287; resemblance to Sina Cactus, 163
Boa rosada. See *Charina trivirgata*
Boidae (pythons and boas), **266–68.**
   See also *Boa constrictor; Charina trivirgata*
Boojum. See *Fouquieria columnaris*
Bowen, Tom, 55, 143, 257
Brown Vinesnake. See *Oxybelis aeneus*
Brush lizards. See *Urasaurus*
Buege, Douglas, 184–85
Bumphead Parrotfish. See *Scarus perrico*
Burgos, Adolfo, iv, 21, 57, 88, 107, 233, 270, plate 9
Burgos, Anita, iv
Burgos, Carmen, iv
Burgos, María, iv
Burrowing owl. See *Athene cunicularia*
Bursera (elephant tree), 71, 91, 95, 120, 134, 145, 146, 147, 151, 203, 250

Cabezona. See *Caretta caretta*
Caborca, 288
Cachora. See *Gambelia wislizenii*

Cachora gris. See *Uta stansburiana*
Cachora matorralera. See *Urosaurus ornatus*
Cachorón. See *Sceloporus clarkii; S. magister*
Caguama cabezona. See *Caretta caretta*
Caguama carrinegra. See *Chelonia mydas*
Cahitan language (Yaqui), 50
Caimán. See *Crocodylus acutus*
Caiman Rock, 46, 47, 78
Callisaurus draconoides (Zebra-tailed Lizard), 32, 73, 75, 99–100, 104, 128, 168, 171, **258–59**, 285, plate 23
Calvino, Italo, 6
Camaleón. See *Phrynosoma solare*
Campo Dolar, 286
Camposano, Roberto, 184, 277
Campo Víboras, 252, 287
Canal del Infiernillo, 46, 78; Comcáac hunting in, 68, 126, 210; declining turtle populations in, 213, 233; density of turtle iime in, 67; Green Sea Turtle in, 179, 197, 198, 240, 243, 245; Hawksbill in, 247; non-Comcáac hunting in, 196; seasnakes in, 276; turtle migration in, 243, 251–52
Canyon Spotted Whiptail Lizard.
   See *Cnemidophorus burti*
Captive breeding, of chuckwallas, 84–85, 219–21, 220, 221, 258
Cardón. See *Pachycereus pringlei*
Caretta caretta (Loggerhead Turtle), 34, 77, 128, 168, 171, **233–34,** plate 6; black market sales of, 187; in Canal del Infiernillo, 233; Comcáac classification of, 114; distinguishing characteristics, 112–13; song about, 234
Carey. See *Eretmochelys imbricata*
Carnegiea gigantea (Saguaro Cactus), 54, 91, 94, 95
Carvings: Comcáac, 1–2, 8, 9, 42, 79, 102, 108, 146–51, 147, 148, 149, 151, 159, plates 15,16,23; non-Seri, 7–8, 149–50
Cascabel del tigre. See *Crotalus tigris*

Case, Ted, 79, 80, 83, 89
Catholicism, influence on Sonoran Indian groups, 51
Central Gulf coast (Sonora): endangered species of, 30–31, 31–33; reptile habitats on, 171–73; reptile species of, 73–75; rich flora and fauna of, xvi, 19, 20, 21–23, 71–72. See also Canal del Infiernillo; Midriff islands
Ceremonies: decline in participation in, 210; girls' puberty, 15–18; Leatherback Turtle, 189–90, 198–99, 210, 250–51. See also Rituals
Charina trivirgata (Rosy Boa), 32, 74, 76, 104, 129, 169, 172, **268,** 287
Chelonia mydas (Green Sea Turtle), 34, 77, 112–13, 128, 141, 168, 171, **234–46,** 283, plates 5,7,11; butchering of, 126, 186, 236; carvings of, 147; ceremonial uses of, 16–18, 143, 234–35, 238; and cleaner wrasse, 156, 167; Comcáac esteem for, 179–81, 235; consumption of, 142, 237; cooyam ("pilgrim/migrant"), 17, 68, 238, 239, 243–44; cooyam caacoj ("large pilgrim/big migrant"), 240, 244; disruption of migration patterns, 252; distinguishing characteristics of, 112–13; as educational model, 218; estivation and hibernation sites, 65; as food, 17, 142, 143; gathering grounds, 67; hunting of, 17, 186–87, 235–36, 240; ipxom haaco iima ("one who is corpulent"), 240, 244; medicinal uses of, 238–39; moosnáapa ("true sea turtle"), 241; moosni áa ("real sea turtle"), 241; moosni ctam hax iima ("sea turtle male with hardly any testes"), 244; moosníctoj ("reddish sea turtle"), 242; moosni hant coit ("sea turtle touching ground"), 245; moosníil ("bluish sea turtle"), 241; moosni

quimoja ("dying sea turtle"), 244; morphological terms for, *114, 239;* quiquíi ("weak/shriveled"), 242–43; size-class distribution in Canal del Infiernillo, *198;* songs about, 240; uses of carapace, 237

Cheloniidae (sea turtles), **233–53,** plates *5,6,7,11,12;* associated with beauty, utility, and danger, *128;* black market sales of, *187;* bladder uses, *144, 144,* 236–37; and Bumphead Parrotfish (*perico*), *157, 167,* 246; butchering of, *126, 186, 236,* plate *13;* carapace uses, *17, 144,* 237–38, 247; ceremonies involving, 16–18, 235, 238; and cleaner wrasse, *156, 167;* Cmique Iitom names for body parts, *114;* Comcáac classification of, *112–13;* decline of, 18–19, 67, *142–43;* depletion during dormancy, 246; disruption of migration patterns, 252; distinguishing characteristics of, *112–13;* egg-laying sites, 68; endangered, *34;* as food, *140–43, 142,* 186, 236, 237, plate *14;* habitats of, *168;* harpooning song, 242; hunting of, *10, 180–81, 186, 235–36,* 240, 245; material uses of, *144–45,* 236–38, 247; meat prized by Mexicans, *186;* medicinal uses, 238–39, 243; morphological terms for, *114, 239;* mythic, 241–42; nonspecific variants in *Chelonia* gene pool, 241–46; as objects of artistic expression, 145–48, *147, 148,* plates *15,16,17,19;* putative Leatherback-Loggerhead hybrid, 252–53; relationships with other species, *166, 228;* species in Sea of Cortés, *77.* See also *Caretta caretta; Chelonia mydas; Eretmochelys imbricata; Lepidochelys olivacea*

Chilomeniscus (sandsnakes), *104*
Chilomeniscus cinctus. See *Chilomeniscus stramineus*

Chilomeniscus punctatissimus (Isla Espíritu Santo Sandsnake), *25, 76*
Chilomeniscus savagei (Isla Cerralvo Sandsnake), *25*
Chilomeniscus stramineus (Bandless Sandsnake), *32, 74, 129, 169, 172,* **269,** 288; disappearance of customs related to, *212,* 269
Chionactis occipitalis (Western Shovelnosed Snake), *74, 162, 169, 172,* **269–70,** 288
Cholludo. See Isla Cholludo
Chuckwalla Nursery Grounds Refuge, 258
Chuckwallas. See *Sauromalus*
Cincuante. See *Pituophis melanoleucus*
Cleaner wrasse. See *Thalassoma lucasanum*
Cmique Iitom language, 3–6; bilingualism, 38; contemporary use of, 35–36; dialect groups, 50; ecological referents for reptiles, *161;* folk taxonomies for reptiles, *101–17, 102, 104, 106, 109, 114;* language isolate, *21;* linguistically encoded knowledge transmitted through traditional activities, 204; loss of, *213;* place-names, 44–49, 52; preservation of, 205, 217–18; pronunciation guide, xiii; reptiles' ability to understand, *113, 130, 146, 192;* Seri dictionary, 44; use of habitat markers in naming reptiles, *161, 162.* See also Comcáac Indians
Cnemidophorus (whiptail lizards), *104, 120, 167*
Cnemidophorus bacatus (Isla San Pedro Nolasco Whiptail), *76*
Cnemidophorus burti (Canyon Spotted Whiptail), *74, 169, 172,* **264**
Cnemidophorus canus (Isla Salsipuedes Whiptail), *24*
Cnemidophorus carmenensis (Isla Carmen Whiptail), *24*

Cnemidophorus catalinensis (Isla Santa Catalina Whiptail), *24*
Cnemidophorus ceralbensis (Isla Cerralvo Whiptail), *24*
Cnemidophorus danheimae (Isla San José Whiptail), *24*
Cnemidophorus espiritensis (Isla Espíritu Santo Whiptail), *24, 76*
Cnemidophorus franciscensis (Isla San Francisco Whiptail), *24*
Cnemidophorus martyris (Isla San Pedro Mártir Whiptail), *24, 76*
Cnemidophorus pictus (Isla Monserrate Whiptail), *24*
Cnemidophorus tigris (Western Whiptail), *74, 76, 129, 169, 172,* **264,** 287
Coachwhips. See *Masticophis estabenensis; M. flagellum; M. slevini*
Cocaznáacöl. See *Arizona elegans; Phyllorhynchus decurtatus; Pituophis melanoleucus; Salvadora hexalepis*
Cocázni. See *Crotalus*
Cocázni cahtxíma. See *Crotalus tigris*
Cocázni quiham. See *Crotalus tigris*
Cocázni sin yeesc. See *Crotalus angelensis; C. estebanensis*
Cocodrilo. See *Crocodylus acutus*
Cocopa Indians. *See* Cucupá
Cocsar (non-Comcáac) hunting of sea turtles, *180*
Cody, Martin, 82, 83
Co-evolution, 28
Cofécöl. *See* Isla San Esteban
Coftj. See *Micruroides euryxanthus*
Coftj caacoj. See *Rhinocheilus lecontei*
Coimaj Caacoj (mythical sea serpent), 41–43, 56–59
Coimaj cmasol. See *Masticophis bilineatus; M. flagellum; M. slevini*
Coimaj coopol. See *Masticophis bilineatus; M. flagellum; M. slevini*
Coimaj coospoj. See *Hypsiglena torquata*
Coimaj cquihöj. See *Masticophis bilineatus; M. flagellum; M. slevini*
Coimaj quiimapxiij. See *Arizona elegans*

Coleonyx (geckoes), *104*

Coleonyx gypsicolus (Isla San Marcos Barefoot Banded Gecko), 24

Coleonyx variegatus (Western Banded Gecko), 32, 73, 76, 128, 169, 172, **262–63,** 286

Collared lizards. See Crotaphytus

Colubridae (colubrid snakes), **268–75.** See also Arizona elegans; Chilomeniscus spp.; Chionactis occipitalis; Hypsiglena spp.; Masticophis spp; Oxybelis aeneus; Phyllorhynchus decurtatus; Pituophis melanoleucus; Rhinocheilus spp.; Salvadora hexalepis; Trimorphodon biscutatus

Comcáac Indians: artistic expressions (see Arts and crafts); belief that reptiles understood Cmique Iitom, 130, 146; bilingualism, 38; comfort around snakes, 122–26; as conservationists, 181–99, 218–23; contemporary homelands, *45;* cultural and linguistic isolation, 28; demographic trends, 36–38, 186; and diabetes, 218; dialect groups, 50; dietary prohibitions, 103, 144, 192, 237; dwelling places, 63–64; historical territory, *46;* hunting of Green Sea Turtle, 186–87, 235–36, 240; ihízitim, 53–56; intellectual property rights, 137–38, 149–50; loss of traditional knowledge, 202–23; material uses for turtles, 144, 236–38; medical uses of reptiles, 138–40, *141;* mestizo population, 36; names for plants, 162–63, *164–65;* names for reptiles, 161, *162;* nutritional reliance upon reptiles, 140–41, 234–35; origin myth, 43–44, 101; as para-ecology consultants, 194–97, 222–23; pets and, 175–77, 212, 248, 260, 261; place-names, 44–49, 52; political activism, 196; as resource managers, 181–84; rights to biological/natural resources, 9, 84, 149, 181–83; role

in plant dispersal, 90, 95–96; role in reptile dispersal, 70–72, 76, *78, 79*–90; sacred sites, 182; and scientific analyses of species dispersal, 84; sense of place, 52–59, 70; taboos, 130, 192–93, 250; understanding of interspecies relationships, *166,* 228; views on reptilian beauty, danger, and utility, 125–26, *128*–29; vision quests, 55, 238, 243, 244. *See also* Cmique Iitom language; Traditional knowledge

Comito, Rafael, iv

Comito, Ricardo Francisco, iv

Comito, Samuel, iv

Commercial fishing. *See* Fishing, commercial

Common Kingsnake. See Lampropeltis getula

Comunidad y Biodiversidad, 195, 197

Conservation/preservation: chuckwalla captive breeding program, 84–85, 219–21, 258; Comcáac and, 9, 34, 181–99, 218–23; of Comcáac language, 205, 217–18; criteria for, 30; ethics of, 11; of ironwood, 7–8, 149–50, 189; para-ecological conservation program, 194–97, 222–23; of sea turtles, 34, 178, 185–86, 188–89, 246, 252; Seri Ethnozoology Education Project, 216–18; and traditional knowledge, 3–4, 38, 215. *See also* Endangered species

Coof. See Sauromalus

Coof coopol. See Sauromalus hispidus

Coof Coopol Iti Ihom. *See* Isla San Lorenzo

Cooyam. See under Chelonia mydas

Cooyam caacoj. See under Chelonia mydas

Coralillo. See Micruroides euryxanthus

Coralillo falso. See Chilomeniscus stramineus; Chionactis occipitalis; False coralsnake; Rhinocheilus lecontei

Coralito, 285

Coralsnake. See Micruroides euryxanthus

Corúa. See Boa constrictor; Coimaj Caacoj; Serpent, spring-dwelling

Cozíxoj. See Coleonyx variegatus; Phyllodactylus xanti; Xantusia vigilis

Crocodiles. See Crocodylus acutus

Crocodylus acutus (River Crocodile), 34, 75, 77, 104, 129, 162, 170, 173, **280–81,** 290; associated with beauty and danger, *129;* of central Gulf coast and midriff islands, 75; Comcáac classification of, 106–7; endangered, *34;* habitats on Sonoran coast and islands, 170, 173; historic presence in northern Mexico, 72, 280–81; in Sea of Cortés, 77

Cross-cultural contact, 23; with O'odham, 36, 49; with peoples of Baja California, 135; with Yaqui, 50; with Yoemem, 36

Crotalus (rattlesnakes), *104;* sharing shelters, 66. See also Viperidae

Crotalus angelensis (Isla Ángel de la Guarda Rattlesnake), 25, **277**

Crotalus atrox (Western Diamondback Rattlesnake), *33,* 74, 76, 129, 169, 172, **277–79,** 289; distinguishing characteristics of, 109; as food, 278; sharing shelters with other reptiles, 65, 66, 278

Crotalus catalinensis (Isla Santa Catalina Rattlesnake), 25

Crotalus cerastes (Sidewinder), *33, 75,* 119–20, 129, 141, 170, 173, **279,** 290; carvings of, *108, 149;* Cmique Iitom names for body parts, *109;* Comcáac classification of, 107

Crotalus enyo (Baja California Rattlesnake), 76

Crotalus estebanensis (Isla San Esteban Rattlesnake), 25, *33, 75, 170, 173,* **277,** 290

Crotalus mitchellii (Speckled Rattlesnake), 25, 76

Crotalus molossus (Black-tailed Rattle-snake), *33, 75, 170, 173,* **277–79,** 290

Crotalus muertensis (Isla El Muerto Rattlesnake), *25*

Crotalus ruber (Red Diamond Rattlesnake), *76,* 290

Crotalus scutulatus (Mohave Rattlesnake), *33, 75, 170, 173,* **277–79,** 290; distinguishing characteristics, *108*

Crotalus tigris (Tiger Rattlesnake), *33, 75, 170, 173,* **280,** 290; Comcáac classification of, *107*

Crotalus tortugensis (Isla Tortuga Rattle-snake), *25*

Crotaphytidae (collared and leopard lizards), **253–54**

Crotaphytus (collared lizards), *104;* Comcáac aversion to, *122*

Crotaphytus dickersonae (Tiburón/ Dickerson's Collared Lizard), *73, 128, 171,* **253,** 284

Crotaphytus insularis (Isla Ángel de la Guarda Collared Lizard), *24, 75*

Crotaphytus nebrius (Sonoran Collared Lizard), *73, 128, 168, 171,* **253,** 284, 286

Ctaamjij. See Crotalus cerastes

Ctamófi. See Callisaurus draconoides

Ctenosaura (spiny-tailed iguanas), *104.* See also Iguanidae

Ctenosaura conspicuosa (Isla San Esteban Spiny-tailed Iguana), *24, 72, 73, 111, 128, 168, 171,* **254–56,** 284, plates *3,4;* Cmique Iitom names for body parts, *111*

Ctenosaura nolascensis (Isla San Pedro Nolasco Spiny-tailed Iguana), *24, 75, 168, 171,* **254–56,** 284

Ctoicj. See Callisaurus draconoides; Uta stansburiana

Ctoixa. See Cnemidophorus burti; C. tigris

Ctoixa iipíil (hatchlings). See Cnemi-dophorus burti; C. tigris

Cubillas, Elvira, iv

Cucupá (Cocopa Indians), 22, 23

Cuernitos. See Crotalus cerastes

Cuija. See Xantusia vigilis

Culebra brillante. See Arizona elegans

Culebra chata. See Salvadora hexalepis

Culebra de la noche. See Hypsiglena torquata

Culebra de los médanos. See Chilomenis-cus stramineus

Culebra látigo. See Masticophis bilineatus

Culebrita. See Phyllorhynchus decurtatus

Cultural dispersal of species, 70–97

Danger: of Desert Iguanas, *127, 128;* reptiles associated with, *125–27, 128–29*

Darwin, Charles, 255

Dátil. See Isla Dátil

Dead Side-blotched Lizard. See Uta lowei

de Bohorques, Lorenzo, 47

Dermochelyidae (leatherback turtles), **249–53**

Dermochelys coriacea (Leatherback Turtle), *34, 77, 112–13, 128, 168, 171, 190, 243,* **249–52,** 284, plates *16,19;* capacity to shed tears, 249; carvings of, *148;* ceremony, 146, 188–89, 198–99, 210, 250–51; Comcáac dietary exclusion of, 103, 250; disruption of migration patterns, 252; distinguishing characteristics of, *112–13;* egg-laying sites, 68, 251; as endangered, 31, 187–88, 251; face-painting designs, 146; historic presence in Sea of Cortés, 146; impact of commercial fishing on, 251; as sacred to Comcáac, 189, 249–50; sale on black market, 187; size of, 249; song about, 198–99, 250; understanding of Cmique Iitom, 20, 103, 130, 146, 250–52

Dermochelys coriacea × Caretta caretta (Leatherback-Loggerhead Hybrid Sea Turtle), **252–53**

Desemboque village, *46,* 97, 147, 156, 209, 216, 283, 284, 285, 286, 287, 288, 289, 290; Cmique Iitom in, 35–36; cultivated plants of, *91–93*

Desert Iguana. See Dipsosaurus dorsalis

Desert Night Lizard. See Xantusia vigilis

Desert Spiny Lizard. See Sceloporus magister

Desert Tortoise. See Gopherus agassizii

Diabetes, 217, 218

Díaz, María del Carmen, iv, *124,* 278

Díaz, Mercedes, iv

Dickerson's Collared Lizard. See Crotaphytus dickersonae

Dinwiddie, William, *48*

Dipsosaurus (desert iguanas), *104.* See also Iguanidae

Dipsosaurus catalinensis (Isla Santa Catalina Desert Iguana), *24*

Dipsosaurus dorsalis (Desert Iguana), *73, 75, 128, 168, 171,* **256,** 284; Comcáac aversion to, 122, 127, 256; danger of, 127, 130; distinguishing characteristics of, *110;* as food, 127; taboos concerning, 130; understanding of Cmique Iitom, 130

Dogs: as pets, 55, 175; use in hunting Desert Tortoises, 69, 209

Dwelling places, human and reptilian. See Iime

Dyes, used in Comcáac art, 146

Ecological knowledge (of indigenous peoples). See Traditional knowledge

Ecological referent, 161

Ecotourism. See Tourism

Eelgrass. See Zostera marina

Elapidae (coralsnakes), **275**

Elephant tree. See Bursera

El Palmar, 287

El Sauzal Arroyo, 284, 286, 288, 290

Endangered languages, 29–30, 35–36

Endangered species, 29–31, 31–33, 34, 35; Comcáac monitoring of, 222; Leatherback Turtle as, 31, 187–88; sea turtles as, 17–18, 67, 142–45, 185–88

Endemic ecological knowledge. *See* Traditional knowledge

Endemic species, 21–23, *24–27*

Ensenada Blanca, 284, 285, 287, 289

Ensenada de la Cruz, 286, 288

Ensenada del Perro, 285, 288, 289, 290

Equilibrium theory, of island biogeography, 81–84

Eretmochelys imbricata (Hawksbill Turtle), *34, 77, 112–13*, 128, *168, 171*, 243, **246–48**, 283; carvings of, *148;* Comcáac classification of, 114; distinguishing characteristics of, *112–13, 113;* ritual of release, 247; uses of carapace, 237–38, 247

Eridaphas marcosensis (Isla San Marcos Nightsnake), *25*

Escorpión de la piedra. See *Crotaphytus dickersonae; C. nebrius*

Escorpión pintado. See *Heloderma suspectum*

Espinosa-Reyna, Alejandrina, 146

Estrella, Efraín, iv

Estrella, Ricardo, iv

Ethnobiology, 4, *5–6*

Ethnoecology, 159–60

Eublepharidae (eyelidded geckoes), **262–63.** See also *Coleonyx gypsicolus; C. variegatus;* Gekkonidae

Eyelidded geckoes. *See* Eublepharidae

Face painting, 146

False coralsnake, 131; Comcáac classification of, 108–9

Felger, Richard, 7, 66, 72, 90, 114, 115, 142, 167, 230, 237–39, 241, 243, 246, 248, 256, 278, 281; "Seri Indian Pharmacopoeia," 135, 140

Félix, Jesús, 257

Félix, Joséfina Ibarra, iv, 160

Félix, Lydia Ibarra, iv, 158

Félix, María, iv, 131, 212

Fish: endangered, *34;* interaction with sea turtles, 156–57

Fishing, commercial: impact on sea turtles, 18–19, 188, 251; for sea turtles, 186, 187; shrimp trawlers, 19, 188, 194, 196, 197, 235; small-scale trawling apparatus (monkey), 196; of totoaba, 87–88; turtle-excluder devices, 188. *See also* Hunting

Flores, Manuelito, iv, 105

Flores, María Ofelia, iv

Folk beliefs. *See* Traditional knowledge

Folklore. *See* Mythology; Traditional knowledge

Folk taxonomies, 101–17, *102;* and classification of sea turtles, *112–13;* and Comcáac animal names, 101–7; and cultural evolution, 115–16; folk varieties, 115; heterogeneity of Comcáac perspectives, 117; and Linnaean taxonomy, 109–10; overclassification, 115; polytypic genera, 103; unaffiliated genera, 103; unique beginner generic term, 101

Food, reptiles as, *127, 140–43, 142,* 236, 237, 278, 279, *plate 12*

Fouquieria columnaris (Boojum), 21, *31,* 119, 249

Gambelia wislizenii (Long-nosed Leopard Lizard), *32, 73, 104, 128, 162, 168, 171,* **254,** 284

Games, 16–17

Geckoes: beliefs about, 121; Comcáac aversion to, 89, 121, 212–13, 263; cultural dispersal of, 72, 80, 88–90; distribution of, 89; as domesticated reptiles, 80. *See also* Eublepharidae; Gekkonidae

Geco. See *Coleonyx variegatus; Phyllodactylus xanti*

Gekkonidae (geckoes), **263–64.** *See also* Eublepharidae; Geckoes; *Phyllodactylus*

Gila Monster. See *Heloderma suspectum*

Global Convention on Biodiversity, 138

Glossy Snake. See *Arizona elegans*

Golfina. See *Lepidochelys olivacea*

Gopher Snake. See *Pituophis melanoleucus*

Gopherus agassizii (Desert Tortoise), *32, 65, 73, 102, 104,* 128, *141,* **227–31,** 283, *plate 9;* abundance in Comcáac territory, 183; archaic name for, 227; beliefs about, 230; burrows of, 65, 66, 69, 228, 278; carvings of, *102;* Cmique Iitom names for body parts of, *106;* consumption of, 140, *143,* 143, 144; consumption of eggs, 211, 229; distribution of, 88–89; as endangered, *32;* forage plants, *166, 170, 173,* 229; habitats of, *168, 171;* hunting of, 69, 191, 209, 230; population of, 228–30; predators of, 229–30; sharing shelters with other reptiles, 66, 228; relationships with other species, *166;* social behavior of, 228–29; song about, 231; taboos concerning, 130, 144, 191–92; understanding of Cmique Iitom, 20, 130, 192–93; use of dogs in hunting of, 69, 209; uses of, 230–31; vocalization during mating, 229

Green, Rayna, 158

Green Sea Turtle. See *Chelonia mydas*

Grismer, L. Lee, 72, 79, 80, 88, 255

Guaymas, 285, 288, 289, 290

Güico. See *Cnemidophorus*

Guirotillo. See *Oxybelis aeneus*

Gulf of California. *See* Sea of Cortés

Haasj. See *Sceloporus*

Hamísj catójoj. See *Oxybelis aeneus*

Hant caxaat. See *Holbrookia maculata*

Hant coáaxoj. See *Phrynosoma solare*

Hantpízl. See *Gambelia wislizenii*

Hant quip. See *Chionactis occipitalis*

Hapéquet camízj. See *Chilomeniscus stramineus*

Haquímet (small lizards), *102, 104, 112,* 150, 259

Harding, Sandra, 173–74
Harmon, David, 14, 28, 30, 35
Hastáacoj. *See* Isla Dátil
Hast coof. *See Crotaphytus dickersonae; C. nebrius*
Hast Heepni. *See* Caiman Rock
Hast Heepni Iti Ihom. *See* San Pedro Nolasco
Hastísil. *See* Isla Cholludo
Hawksbill Turtle. *See Eretmochelys imbricata*
Hayden, Julian, 176–77
Healing practices. *See* Medical treatments
Heepni. *See Ctenosaura conspicuosa; C. nolascensis*
Hehe iti cooscl. *See Urosaurus ornatus*
Heloderma suspectum (Gila Monster), *32, 65, 74, 104, 129, 141, 169, 172,* **265–66,** 287, *plate 22;* carvings of, *151;* Comcáac tolerance for, 130; as food, 142; infrequency of attacks on humans, 131; medicinal use of, 266; sharing shelters with other reptiles, 65, *66,* 228
Herbs, medicinal: Comcáac intellectual rights to, 138; for ray stings, 132; for snake bites, 135, *136, 137,* 211; for scorpion stings, 132–33; for seasnake bites, 276. *See also* Plants
Hermosillo, 286
Herpetofauna. *See* Reptiles; *names of particular species*
Herrera, Genaro, iv, 29, 34, *194,* 222, 223
Herrera, Juanita, iv, 275
Hia c-eḍ O'odham. *See* O'odham Indians (Papago-Pima)
Hills, Jim "Santiago Loco," 7–8
Hoeffer, Gabriel, iv, 19, 189
Hoeffer, Raquel, iv
Hogan, Linda, 40
Hokan language superfamily, 3
Holbrookia maculata (Lesser Earless Lizard), *73, 168,* **281,** *285*

Horned lizard. *See Phrynosoma solare*
Hunn, Gene, 112, 115
Hunting: of chuckwallas, 191; of Desert Tortoise, 69, *125,* 191, 209, 231; of sea turtles, *126,* 180–81, 210, 235–36, 240, 245, *plates 11,12,13,20;* of snakes, *125;* taboos protecting animals, 192; use of dogs in, 69, 209. *See also* Fishing, commercial
Hydrophiidae (seasnakes), **276**
Hypsiglena gularis (Isla Partida Norte Nightsnake), *25*
Hypsiglena torquata (Nightsnake), *33, 74, 76,* 129, *169, 172,* **270,** 288

Iguana. *See Ctenosaura; Sauromalus*
Iguanidae (iguanas and chuckwallas), **254–58;** cultural dispersal of, 72, 76–77, *78, 79;* as food, 140; nesting and congregating places, 65. *See also Ctenosaura; Dipsosaurus; Sauromalus*
Ihízitim, 53–56; defined, 54
Iicj Icóoz. *See* Isla San Pedro Mártir
Iime ("home," "resting place," "shelter"): and Comcáac sense of place, 70; defined, 63; for desert reptiles, 65, *66,* 69; human, 63–65; for sea turtles (moosni), 66–68
Indigenous knowledge. *See* Traditional knowledge
Indigenous medical practices. *See* Medical treatments; Traditional knowledge
Indigenous peoples: epistemic privileges of, 184–85; essentializing of, 184; as "first ecologists," 157, 183; hunting practices of, 191
Indigenous science. *See* Traditional knowledge
Initiation ceremonies. *See* Ceremonies; Rituals
Instituto Nacional de Ecología, 181
Interspecific relationships: between

Boa Constrictor and Sina Cactus, 163; between Brown Vinesnake and Ashy Limberbush, 167; Comcáac knowledge of, 157–59, *166;* between Green Sea Turtle and cleaner wrasse, 155–57, 167; between pets and people, 175–77; reflected in language, 161, *162, 164–65;* sharing shelters, 65–66, 228
Ipxom haaco iima. *See under Chelonia mydas*
Ironwood. *See Olneya tesota*
Ironwood Alliance, 8, 149–50, *150,* 197
Isla Alcatraz, *46;* chuckwallas on, 63, *77, 78, 85, 86, 87, 88,* 191, *plates 1,2;* leaf-toed geckoes on, 80, 89; prickly pear on, 90; reptile species on, 71, *73–75, 285, 286, 287*
Isla Ángel de la Guarda, *46;* chuckwallas on, *77,* 105, 257; fauna of, 24–26; flora of, *27;* leaf-toed geckoes on, 89; rattlesnakes on, 20, 277; reptile species on, 24–25, *75–76,* 283, 285, 286, 287, 288, 289, 290
Isla Ángel de la Guarda Collared Lizard. *See Crotaphytus insularis*
Isla Ángel de la Guarda Rattlesnake. *See Crotalus angelensis*
Isla Carmen Whiptail. *See Cnemidophorus carmenensis*
Isla Cerralvo Long-nosed Snake. *See Rhinocheilus etheridgei*
Isla Cerralvo Sandsnake. *See Chilomeniscus savagei*
Isla Cerralvo Spiny Lizard. *See Sceloporus grandaevus*
Isla Cerralvo Whiptail. *See Cnemidophorus ceralbensis*
Isla Cholludo, *46;* chuckwallas on, *78;* fauna of, 24–26; and leaf-toed gecko, 89; reptile species on, 71, *73–75,* 287; spiny-tailed iguana on, 72, *76–77, 78,* 88, 254, 255–56

Isla Dátil, *46, 78;* and Desert Tortoise, 88–89; reptile species on, *71, 73–75,* 283, 286, 288, 289; and spiny-tailed iguana, 76, 88

Isla El Muerto Rattlesnake. See *Crotalus muertensis*

Isla Espíritu Santo Striped Racer. See *Masticophis barbouri*

Isla Espíritu Santo Whiptail. See *Cnemidophorus espiritensis*

Isla Monserrate Whiptail. See *Cnemidophorus pictus*

Isla Partida Norte Leaf-toed Gecko. See *Phyllodactylus partidus*

Isla Salsipuedes Whiptail. See *Cnemidophorus canus*

Isla San Esteban, *46;* breeding of chuckwallas from, 84–85, 219–20; chuckwallas on, *19,* 20, 62, *78,* 83, *86, 87;* Comcáac living/hunting on, *51, 52, 55,* 143, 191; cultural importance of, 183; endangered species of, *31–33;* fauna of, *24–26;* flora of, *27;* leaf-toed gecko on, 80, 89; Organpipe Cactus on, *95;* protection of, 181; rattlesnakes on, 20; reptile species on, *73–75,* 284, 285, 286, 287, 288, 289, 290; spiny-tailed iguana on, 72, *78,* 88, 110, 254–56, *plates 3,4;* taboos protecting animals on, 192; unique coexistence of reptile species, 72

Isla San Esteban Chuckwalla. See *Sauromalus varius*

Isla San Esteban Coachwhip. See *Masticophis slevini*

Isla San Esteban Rattlesnake. See *Crotalus estebanensis*

Isla San Esteban Spiny-tailed Iguana. See *Ctenosaura conspicuosa*

Isla San Francisco Whiptail. See *Cnemidophorus franciscensis*

Isla San José Whiptail. See *Cnemidophorus danheimae*

Isla San Marcos Barefoot Banded Gecko. See *Coleonyx gypsicolus*

Isla San Marcos Nightsnake. See *Eridaphas marcosensis*

Isla San Pedro Mártir, *46;* fauna of, *24–26;* reptile species on, *72, 76,* 284, 287, 289

Isla San Pedro Mártir Side-blotched Lizard. See *Uta palmeri*

Isla San Pedro Mártir Whiptail. See *Cnemidophorus martyris*

Isla San Pedro Nolasco, *46;* fauna of, *24–26;* flora of, *27;* reptile species on, *72, 75–76,* 285, 289; spiny-tailed iguanas on, *72, 78,* 88

Isla San Pedro Nolasco Side-blotched Lizard. See *Uta nolascensis*

Isla San Pedro Nolasco Spiny-tailed Iguana. See *Ctenosaura nolascensis*

Isla Santa Catalina Chuckwalla. See *Sauromalus klauberi*

Isla Santa Catalina Desert Iguana. See *Dipsosaurus catalinensis*

Isla Santa Catalina Leaf-toed Gecko. See *Phyllodactylus bugastrolepsis*

Isla Santa Catalina Rattlesnake. See *Crotalus catalinensis*

Isla Santa Catalina Side-blotched Lizard. See *Uta squamata*

Isla Santa Catalina Spiny Lizard. See *Sceloporus lineatulus*

Isla Santa Catalina Whiptail. See *Cnemidophorus catalinensis*

Isla Santa Cruz Spiny Lizard. See *Sceloporus angustus*

Islas San Lorenzo Norte and Sur, *46;* chuckwallas on, *78, 86, 87,* 105, 143, 257; fauna of, *24–26;* flora of, *27;* leaf-toed gecko on, 80, 89; rattlesnakes on, 20; reptile species on, *72, 75–76;* river crocodiles on, 281

Isla Tiburón, *46, 78;* Comcáac living on, *52, 53, 65, 71, 107, 116, 125,* *126, 176;* and Comcáac origin story, 44; commericial fishing near, 196; cultural importance of, 183; Desert Tortoise on, 20, 69, 144, 170, 228, 230; fauna on, *25–26;* flora on, *27;* hunting on, 184, 192, *plates 13,14;* leaf-toed gecko on, 80, 88, 89; para-ecologists on, 194; place-name, 47; protection of, 181, 184, 195; reptile species on, *72, 75–76,* 283–90; sacred sites on, 182; and sea turtles, 67, 68, 179; spiny-tailed iguana on, 76, *78,* 88, 110, 254–56

Isla Tortuga Rattlesnake. See *Crotalus tortugensis*

Isopods, 120, 279

Ivanyi, Craig, 65, 85, 100, 264, 268, 272, 274

Jabalina. See *Caretta caretta*

Jatropha cinerea (Ashy Limberbush), *166, 167,* 272

Jewelry (necklaces, rings), 145; from Gila Monster, 266; from Glossy Snake, 268; from Gopher Snake, 273; from rattlesnake, 124, *145,* 209, 277, 279; from whipsnake, 271. *See also* Arts and crafts

Kingsnakes. See *Lampropeltis catalinensis; L. getula*

Kinosternidae (mud and musk turtles), **232–33**

Kinosternon (mud turtles), 104, 162, **232–33**

Kinosternon flavescens (Yellow Mud Turtle), *73, 168,* **232–33,** 283; habitat, *171*

Kinosternon sonoriense (Sonoran Mud Turtle), *73, 128, 162, 166, 168,* **232–33;** as food, 143; habitat, *171*

Kino Viejo, 61

Kloppenburg, Jack, 84

Lagartija. See *Urosaurus ornatus*
Lagartija de cercos. See *Uta stansburiana*
Lagartija de collar. See *Crotaphytus dickersonae; C. nebrius*
Lagartija de leopardo. See *Gambelia wislizenii*
Lampropeltis catalinensis (Isla Catalina Kingsnake), *25*
Lampropeltis getula (Common Kingsnake), *74, 76, 169, 172, plate 10*
Landmarks, reptile shelters as, *70*
Language, indigenous: as endangered, 29–30, 35–36; evolution of, *14*; knowledge embedded in, 22–29, 38, 204–5, 221; loss of, 200, 204–5. *See also* Cmique Iitom language
Las Cachanillas (indigenous inhabitants of Baja California), *55*
Las Cuevitas, *285*
Laúd. See *Dermochelys coriacea*
Leatherback-Loggerhead Hybrid Sea Turtle. See *Dermochelys coriacea × Caretta caretta*
Leatherback Turtle. See *Dermochelys coriacea*
Lepidochelys olivacea (Pacific [Olive] Ridley Sea Turtle), *34, 77, 112–13, 128, 168, 171*, 243, **248–49**, 283; carvings of, *148*; distinguishing characteristics of, *112–13*; egg-laying sites of, 68; as pets, 248
Lesser Earless Lizard. See *Holbrookia maculata*
Lichanura. See *Charina trivirgata*
Life-form categories, *101–3, 102*
Lizards, **253–66;** associated with beauty, utility, and danger, *128–29;* endangered, *32;* habitats of, *168–69, 171–72;* as pets, 175–77; species on central Gulf coast and midriff islands, *24, 73–74, 75–76. See also names of particular species*
Local knowledge. *See* Traditional knowledge

Loggerhead Turtle. See *Caretta caretta*
Long-nosed Leopard Lizard. See *Gambelia wislizenii*
Long-nosed Snake. See *Rhinocheilus lecontei*
Long-tailed Brush Lizard. See *Urosaurus graciosus*
López, Alfredo, iv, *19*, 43, 62, 87, 107, 196, 201, 220, *221*, 232, 280–81; on Comcáac dispersal of chuckwallas, 79; on Desert Tortoise, 69, *193;* on Glossy Snakes, 268; on hybrid sea turtle, 252; on illegal fishing, 197; on places known by Comcáac, 49–50; on plant dispersal and Birthing Place, 95; on turtle hunting, 180–81; on whipsnakes, 271
López, Antonio, iv, *142*, 256
López, David, iv
López, Emilio Anacarsis, iv
López, Eva, *167*, 264
López, Guadalupe, iv, *66, 100, 170*, 228, 236, 259, 278; on Canal del Infiernillo sea turtles, 245–46; on Leatherback Turtles, 103, 250–52; on mythic sea turtle, *241, 242;* on Nightsnakes, 270; on sea turtle and fish interaction, *156–57, 167;* on sea turtle shelters, *65, 67–68*
López, José Luis, iv
López Barnet, Ramón, iv
López Flores, Ramón, iv, *52*, 274
Lyresnake. See *Trimorphodon biscutatus*

Magayan, José Antonio "Delfino," 267
Malkin, Borys, *90, 109*, 100, 127, *142*, 229, 254, 255, 256, 259, 264, 274, 275
Mammals: endangered, *33, 34;* endemic to midriff islands, *25–26*
Marine reptiles: environmental threats to, 194; historically in Sea of Cortés, 77; shrimp trawlers' effects on, 194;

species in Comcáac territory, *71. See also* Reptiles
Marlett, Stephen: Seri dictionary, 44, 54
Martínez, Jesús Alberto, iv
Martínez, José Luis, iv
Martínez, Solorio, iv, *151*
Martínez, Victor, iv
Masticophis (coachwhips and racers), *104*, 228, **270–71**
Masticophis barbouri (Isla Espíritu Santo Striped Racer), *25*, 76
Masticophis bilineatus (Sonoran Whipsnake), *74, 76, 129, 169, 172*, **270–71**, 288
Masticophis estebanensis (Isla San Esteban Coachwhip), *25, 33, 74, 169, 172*, **270–71**, 289
Masticophis flagellum (Sonoran Coachwhip), *33, 74, 104, 169, 172*, **270–71**, 288
Masticophis slevini (Isla San Esteban Coachwhip), *25, 32, 74, 169, 172*, **270–71**, 289
Matpayóo. See *Callisaurus draconoides; Uta stansburiana*
Maxáa. See *Charina trivirgata; Hypsiglena torquata; Masticophis bilineatus; M. flagellum; M. slevini; Salvadora hexalepis*
Mayo Indians. *See* Yoemem Indians
McGee, W. J., 36, 37, 47, *48*, 63, 116, 140, 142, 234
Medical treatments: indigenous, 38, 131–37, *134, 136, 141*, 205, 238–39, 243, 266, 278, *plate 18;* plants used for reptile bites, *132–37, 134, 136, 205, plate 18;* snake oil, *138*, 138–39, 278; using rattlesnakes, *138*, 138–39, 278–79. *See also* Traditional knowledge
Medicine shows, *139*
Mellado, Alberto, iv
Menchú, Rigoberta, 30
Méndez, Alfonso, iv, 228, 276

Méndez, Elena Gastelum, iv
Méndez, Estela, iv
Méndez, Moisés, iv, 9
Méndez, Oswaldo, 228
Méndez, Rodrigo, iv
Méndez, Rosita, iv
Mesquite. See *Prosopis*
Mexican Government, ban on sale
  of sea turtles, 67
Mexico: endemic species, 22–23;
  indigenous inhabitants, 22; unique
  languages, 22
Meyo. See *Dipsosaurus dorsalis*
Micro-areal endemics, 22–23
Micruroides (coralsnakes), *104*
Micruroides euryxanthus (Western/Sonoran
  Coralsnake), *19, 33, 74, 104, 129,
  169, 172,* **275,** *289, plate 24;* Com-
  cáac classification of, 107; Comcáac
  tolerance for, 130–31; venom of,
  131
Midriff islands: cultural dispersal of
  iguanids between, *78;* endemic
  fauna, *24–26;* endemic flora, *27;*
  reptiles nearest Comcáac villages,
  73–75; reptile species, *75–76. See
  also* Isla Alcatraz; Isla Cholludo;
  Isla Dátil; Isla San Esteban; Isla San
  Lorenzo; Isla San Pedro Mártir; Isla
  San Pedro Nolasco; Isla Tiburón
Molina, Anabertha, iv
Molina, Ernesto, iv, 121, 229, 256,
  257; on chuckwallas' scarcity, 191;
  on Comcáac territory extent, 51;
  on Desert Tortoise depletion, 193;
  on Gila Monsters, 266; on horned
  lizards, 130, 260; and Loggerhead
  song, 234; on reptiles, 20; and
  sacred sites on Isla Tiburon, 182;
  on sea turtles, 17, 34, 167; on wildlife protection, 216
Molina, Francisco, iv
Molina, José Guadalupe, iv
Molina, Malca, iv
Molina, María Félix, iv

Molina, Raúl, iv
Molina, Roberto Carlos, iv, 146
Molina, Saúl, iv
Monrroy, Martina, iv
Montaño, Jesús, iv
Montaño, Lydia, iv
Montecinos, Camila, 5
Monti, Laurie, 217–18, 267
Moosnáapa. See under *Chelonia mydas*
Moosni, *104.* See also *Chelonia mydas*
Moosni áa. See under *Chelonia mydas*
Moosni ctam hax iima. See under
  Chelonia mydas
Moosníctoj. See under *Chelonia mydas*
Moosni hant coit. See under *Chelonia
  mydas*
Moosni iime. *See under* Iime
Moosníil. See under *Chelonia mydas*
Moosni ilítcoj caacöl. See *Caretta caretta*
Moosni otác. See *Lepidochelys olivacea*
Moosnípol. See *Dermochelys coriacea*
Moosnípol xpeyo (quimozíti). See
  Dermochelys coriacea × Caretta caretta
Moosni quimoja. See under *Chelonia
  mydas*
Moosni quipáacalc. See *Eretmochelys
  imbricata*
Moosni sipoj. See *Eretmochelys imbricata*
Morales, Arturo, iv
Morales, Cleotilde, iv, 247
Morales, David, iv, 140
Morales, Fernando, iv
Morales, Humberto, iv
Morales, Jesús, 143
Morales, Roberto, iv
Moreno, Armida Patricia, iv, 203
Moreno, José Juan, iv, 175, 234, 255,
  258
Moreno, María del Carmen, iv
Moreno, Patricia, iv
Moreno family, 255
Moser, Becky, 7, 38, 54, 66, 90, 114,
  115, 142, 167, 230, 237–39 passim,
  241, 243, 246, 248, 257, 278; "Seri
  Indian Pharmacopoeia," 135, 140

Mud and musk turtles. *See*
  Kinosternidae
Mythology: of earthquake-causing
  serpent, 281; of giant sea serpent,
  41–43, 56–59; origin myth, 43–44,
  101, 238; of Regal Horned Lizard,
  201–2; spring-dwelling serpent,
  56–58

Native peoples. *See* Indigenous peoples
Natural resources: cultural assess-
  ments of, 122; management of
  (*see* Conservation/protection)
Necklaces, 17; from Gila Monster,
  266; from Glossy Snake, 268; from
  Gopher Snake, 273; from rattlesnake,
  124, 145, 209, 279; from whip-
  snake, 271. *See also* Jewelry
Neotoma (packrats), *26, 66, 160,* 228,
  278
Neotoma albigula (White-throated
  Packrat), 228
Night lizards. See *Xantusia; Xantusiidae*
Nightsnake. See *Eridaphas marcosensis;
  Hypsiglena gularis; H. torquata*
Norris, Ken, 76, 85
Northern Arizona University, 194–95,
  216
Northern Chuckwalla. See *Sauromalus
  obesus*

Olneya tesota (Ironwood), *31;* Comáac
  conservation of, 7–8, 189; as
  habitat, 142; overuse of by non-Seri
  carvers, 7–8, 149–50; use in traditional carving, 8, 146–49, *plates
  20,23*
O'odham Indians (Papago-Pima), 22,
  23, 49, 234, 281; Catholicism's
  influence on, 51; Comcáac contact
  with, 36, 49; Desert Iguanas as food
  for, 127; rattlesnakes and, 121, 122;
  and "staying sickness," 130, 260
Opata Indians (Eudeve), Catholicism's
  influence on, 51

Ophidiophobia (dread of snakes), 122–23, 127, 152, 215

Opuntia bigelovii (Teddy Bear Cholla), 249

Opuntia engelmanii (Prickly Pear), 95, 222

Organpipe Cactus. See Stenocereus thurberi

Origin myth, 43–44, 101, 238

Ornamentation: face painting, 146; rattlesnakes used for, 145, 279. See also Arts and crafts

Ortega, José Guadalupe, iv

Oxybelis aeneus (Brown Vinesnake), 74, 104, 166, 167, 169, 172, **272**

Paaza. See Heloderma suspectum

Pachycereus pringlei (Cardón), 50, 95, 120

Pacific (Olive) Ridley Sea Turtle. See Lepidochelys olivacea

Packrats. See Neotoma

Palo Fierro, 286, 288

Papago-Pima Indians. See O'odham Indians

Para-ecologists: Comcáac as, 194–97, 222–23; role in preventing poaching, 195

Patos, 285

Pelamis platurus (Yellow-bellied Seasnake), 74, 77, 104, 129, 162, 169, 172, **276**, 289; Comcáac classification of, 107; infrequency of attacks on humans, 131; venom of, 131

Peninsular Leaf-toed Gecko. See Phyllodactylus xanti

Perales, Héctor, iv

Perales, Ramón, iv, 68, 114, 170, 202; on Hawksbill Turtle, 247; on Leatherback Turtle, 252; on sea turtles, 233, 243–44, 247

Perico. See Eretmochelys imbricata; Scarus perrico

Perrito. See Callisaurus draconoides

Petrosaurus slevini (Slevin's Banded Rock Lizard), 24

Pets, 175–77; dogs as, 175; reptiles as, 123, 175–77, 207, 212, 214, 248–49, 260

Phrynosoma solare (Regal Horned Lizard), 2, 73, 104, 128, 162, 168, 171, **259–61**, 285; carvings of, 2, 203; classified as human, 201; as motif in basketry, 203; in mythology, 201–2; nesting and congregating places, 65; as pets, 175, 212, 260; song/rhymes about, 201, 260; and "staying sickness," 130, 260; taboos concerning, 130, 260; understanding of Cmique Iitom, 130

Phrynosomatidae (sand, rock, spiny, horned, brush, and side-blotched lizards), **258–62**

Phyllodactylus (leaf-toed geckoes), 104

Phyllodactylus bugastrolepsis (Isla Santa Catalina Leaf-toed Gecko), 24

Phyllodactylus homolepidurus (Sonoran Leaf-toed Gecko), 73, 76

Phyllodactylus partidus (Isla Partida Norte Leaf-toed Gecko), 24, 76

Phyllodactylus unctus (San Lucan Leaf-toed Gecko), 76

Phyllodactylus xanti (Peninsular Leaf-toed Gecko), 74, 76, 128, 169, 172, **263–64**, 287

Phyllorhynchus decurtatus (Spotted Leaf-nosed Snake), **272–73**, 289

Piebald Chuckwalla. See Sauromalus varius

Pitahaya Agria. See Stenocereus gummosus

Pitahaya Dulce. See Stenocereus thurberi

Pituophis melanoleucus (Gopher Snake), 74, 104, 129, 169, 172, **273**, 289

Plants: Comcáac names for, 162–63; Comcáac overclassification of, 115–16; conservation of, 222; cultivated, 91–95, 96; for curing reptile bites, 132–37, 134, 136; diversity of, 20–23; eaten by tortoises, 166, 170, 173, 227, 229; endangered, 31; endemic to Sea of Cortés, 27; medicinal, 132–37, 134, 136, 205, plate 18;

and reptile names, 162–63, 164–65, 169; traditional knowledge of, 22–23, 28–29, 160–61, 208. See also Algae; Herbs, medicinal; names of individual plants

Porohui. See Dipsosaurus dorsalis

Pozo Coyote, 27, 285

Preservation. See Conservation/preservation

Prickly Pear. See Opuntia engelmanii

Prosopis (mesquite), 70, 92, 94, 95, 142, 160, 218, 222, 271, 274

Puberty rites, 15–18, 235, 243

Puerto Libertad, 284, 286, 289, 290

Puerto Lobos, 288

Punta Chueca village, 1, 7, 15, 97, 147, 209, 216, 283, 288; Cmique Iitom in, 35–36; cultivated plants in, 93–95

Punta Cirio, 285

Punta Hona, 283

Punta Sargento, 78, 258, 286

Pythons. See Boidae

Quammen, David, 60

Quiquíi. See under Chelonia mydas

Racers. See Masticophis

Rancho Noche, 289

Rapoport, Eduardo, 21

Rattlesnake oil. See Snake oil

Rattlesnakes. See Crotalus; Viperidae

Red Diamond Rattlesnake. See Crotalus ruber

Reddish sea turtle. See under Chelonia mydas

Regal Horned Lizard. See Phrynosoma solare

Reptile bites, 131–37; diagnosis and treatment of, 136; plants used for curing, 132–37, 134, 136

Reptile dispersal, Comcáac role in, 70–72, 76, 79–83

Reptile habitats: recorded by Seri consultants, 171–73; recorded by western scientists, 168–70

Reptiles: beliefs concerning danger of, 122; Comcáac classification of, 101–17, *104;* Comcáac names for, 161, *162;* conservation of, 194–96, 221; cultural dispersal of, 70–97; endangered species, *32–33, 34;* habitats of, *71–72, 168–73;* medical uses of, *138–40, 141, 278–79;* nesting and congregating places, 65–66, *66 (see also* iime); as nutrition sources, 140–44; as objects of artistic expression, 145–52; relationships with other species, *166,* 228; species on central Gulf coast and midriff islands, *73–75, 75–76;* symbolic significance of, *153;* taboos concerning, 130, 144; threats to, 194; understanding of Cmique Iitom, 130, 146. *See also* Lizards; Marine reptiles; Snakes; *names of particular species*

Resource management and conservation. *See* Conservation/preservation

Rhinocheilus etheridgei (Isla Cerralvo Long-nosed Snake), *25*

Rhinocheilus lecontei (Long-nosed Snake), *74, 169, 172,* **273–74,** 289

Richardson, Maurice, 80, 90, 131

Ricketts, Edward F., 4, 98

Rituals: Bandless Sandsnake in, 212, 269; Birthing Place, *54;* in Hawks-bill Turtle release, 247; for protection from sea serpent, 42–43; puberty, 15–18, 235, 243; sea turtles in, 18, 238, 250–51; shamanic purification, 130; sharing of food, 237; vision quests, *55,* 238, 243, 244. *See also* Ceremonies

River Crocodile. *See Crocodylus acutus*

Robles, Antonio, iv, 9, 222

Robles, Cornelio, iv, 221–22, 271

Robles, Enrique, iv

Robles, Israel, iv

Robles, Josuë, iv

Robles Barnet family, *149*

Rock lizards. *See* Phrynosomatidae

Rojo, Jesús, iv, 3, *175,* 198, 232, 271, *plate 8;* on Comcáac homeland, *53;* on cultural changes, 143, 204, 208; on hunting Desert Tortoise, 192; on Leatherback Turtle, 146, 251; on Pacific (Olive) Ridley Sea Turtle, 248; song of mythic bluish sea turtle, 241, 242; on tortoise forage, *173*

Rojo, Norma Alicia, iv

Romero, Felipe, iv

Romero, Francisco "Ruben," iv

Romero, Humberto, iv

Romero, Magdalena, iv

Romero, Manuel, iv

Romero, Pedro, iv, 9, 36, 182–85, 213, 216, 223

Romero, Rafaela, iv

Romero, Roberto, iv

Romero Astorga, Pedro, iv

Rosenberg, Janice, 216, 259, 260, 263, 278

Rosendo, Alfonso, iv

Rosy Boa. *See Charina trivirgata*

Sacred ecology. *See* Traditional knowledge

Sacred sites, 182

Saguaro Cactus. *See Carnegiea gigantea*

Salamanquesa. *See Phyllodactylus xanti; Xantusia vigilis*

Salamanquesa de franjas. *See Coleonyx variegatus*

Salvadora hexalepis (Western Patch-nosed Snake), *74, 76, 104, 169, 172,* **274,** 289

Sand lizards. *See* Phrynosomatidae

Sandsnake. *See Chilomeniscus stramineus*

San Esteban. *See* Isla San Esteban

San Lorenzo Norte y Sur. *See* Islas San Lorenzo Norte and Sur

San Pedro Mártir. *See* Isla San Pedro Mártir

Sauer, Carl O., 122

Sauer, Jonathan, 81–82

Sauromalus (chuckwallas), 61–63, *104,* 110, **257–58;** captive breeding program, 84–85, 219–21, *220, 221,* 258; captured vs. captive-bred, 86–87; carvings of, *77;* Cmique Iitom names for body parts of, *111;* Comcáac influence on varieties of, 63; conservation of, 258; cultural dispersal of, 72, 76–77, 78, 85–88, *plates 1,2;* diet of, 219; as food, 140, *142,* 143; hunting of, 191; hybrids, *73,* 86–87, *plates 1,2;* incipient domestication of, 86; on Isla San Esteban, 72; medicinal use of, *141;* nesting and congregating places, 65

Sauromalus ater. See Sauromalus obesus

Sauromalus hispidus (Spiny [Black] Chuckwalla), *24, 32, 75, 77,* 86, *128, 168, 171,* **257–58,** *285;* as source of food, *143*

Sauromalus klauberi (Isla Santa Catalina Chuckwalla), *24*

Sauromalus obesus (Northern Chuckwalla), *32, 73, 75, 128, 168, 171,* **257–58,** *285*

Sauromalus slevini (Slevin's Chuckwalla), *24*

Sauromalus varius (Piebald Chuckwalla/ Isla San Esteban Chuckwalla), *19, 24, 32, 73, 75, 77,* 84–87, *128, 168, 171,* **257–58,** *285, plate 21*

Scarus perrico (Bumphead Parrotfish), 246

Sceloporus (spiny lizards), *104*

Sceloporus angustus (Isla Santa Cruz Spiny Lizard), *24*

Sceloporus clarkii (Clark's Spiny Lizard), *73, 75, 168, 171,* **261,** *285*

Sceloporus grandaevus (Isla Cerralvo Spiny Lizard), *24*

Sceloporus lineatulus (Isla Santa Catalina Spiny Lizard), *24*

Sceloporus magister (Desert Spiny Lizard), *65, 73, 128, 168, 171,* 228, **261,** *286;* sharing shelters with other reptiles, *66*

Science, expanded definition of, 158
Sea of Cortés (Gulf of California), 45, 46, 78; geography of, 60; historic presence of marine reptiles in, 77; threatened marine fauna of, 17, 34. See also Canal del Infiernillo; Central Gulf coast (Sonora); Midriff islands
Sea serpent. See Coimaj Caacoj
Seasnakes. See Pelamus platurus
Sea turtles. See Cheloniidae; Dermochelys coriacea
Seminoff, Jeff, 246, 252
Seri dictionary, 44
Seri Ethnozoology Education Project, 216–18, 217
Seri Indians. See Comcáac Indians
Seris Tepoques (offshoot of Comcáac Indians), 49
Serpent, spring-dwelling, 56–58. See also Coimaj Caacoj
Shamanic purification, of psychosomatic trauma, 130
Shrimp trawlers: Comcáac political activism and, 196; control of, 197; destruction of sea turtles and habitats, 188; effects on marine reptiles, 194; impact on sea turtles, 18–19, 235. See also Fishing, commercial
"Shriveled" sea turtle. See under Chelonia mydas
Side-blotched Lizard. See Uta stansburiana
Sidewinder. See Crotalus cerastes
Sierra Seri, 282
Siete filos. See Dermochelys coriacea
Siete filos híbrido con carey. See Dermochelys coriacea × Caretta caretta
Sina Cactus. See Stenocereus alamosensis
Slevin's Banded Rock Lizard. See Petrosaurus slevini
Slevin's Chuckwalla. See Sauromalus slevini
Smith, William Neill, 54, 145, 237, 238
Snakebites: decline in use of traditional remedies for, 211; diagnosis and treatment of, 131–38, 134, 136

Snake oil: huckersterism and, 138, 139; medicinal uses of, 138, 138–39, 278
Snakes, 266–80; associated with beauty, utility, and danger, 125–27, 129; endangered, 32–33; fear of, 122–23, 127, 152, 215; habitats of, 169–70, 172–73; in Sea of Cortés, 77; sharing shelters with other reptiles, 66, 228; species on central Gulf coast and midriff islands, 25, 73–74, 76; symbolic significance of, 153. See also names of particular species
Songs: about chuckwallas, 257; about coachwhip snakes, 271; about collared lizards, 253; as conveyors of traditional knowledge, 52; about Desert Iguana, 256; about Desert Tortoise, 231; about Green Sea Turtle, 240, 242; about Leatherback Turtle, 198–99, 250; about Loggerhead Turtle, 234; about mud turtle, 232–33; preservation of, 216–18; about rattlesnakes, 279; about Regal Horned Lizard, 201, 260; about whiptail lizards, 264; about Zebratailed Lizard, 259
Sonoran Coachwhip. See Masticophis flagellum
Sonoran Collared Lizard. See Crotaphytus nebrius
Sonoran Coralsnake. See Micruroides euryxanthus
Sonoran Desert (coastal region). See Central Gulf coast (Sonora)
Sonoran Leaf-toed Gecko. See Phyllodactylus homolepidurus
Sonoran Mud Turtle. See Kinosternon sonoriense
Sonoran Whipsnake. See Masticophis bilineatus
Sosni. See Isla Alcatraz
Spanish language, as lingua franca, 23
Speckled Rattlesnake. See Crotalus mitchellii
Spicer, Edward, 281

Spiny (Black) Chuckwalla. See Sauromalus hispidus
Spiny lizards. See Sceloporus
Spiny-tailed Iguana. See Ctenosaura conspicuosa; C. nolascensis
Spiritual traditions, impact on species survival, 183
Spotted Leaf-nosed Snake. See Phyllorhynchus decurtatus
Stenocereus alamosensis (Sina Cactus), resemblance to Boa Constrictor, 163, 164, 267
Stenocereus gummosus (Pitahaya Agria), 54, 91, 94, 95, 258
Stenocereus thurberi (Organpipe Cactus), 50, 94, 95, 120
Steinbeck, John, 4, 98
Swollen-nosed Side-blotched Lizard. See Uta tumidarostra

Taboos: associated with reptiles, 130, 250, 260, 265; ecological function of, 193; impact on reptile species' depletions, 192–94
Tahéjöc. See Isla Tiburón
Tastiota, 285, 288, 289, 290
Tecomate, 285
Teddy Bear Cholla. See Opuntia bigelovii
Teiidae (whiptail lizards), 24, 76, 120, 264; song about, 167. See also Cnemidophorus
Teiwes, Helga, 9
Terralingua, 28
Testudinidae (land tortoises), 227–31. See also Gopherus agassizii
Thalassoma lucasanum (Cortés rainbow wrasse), interdependence with sea turtles, 156, 167
"Thick" description, 5
Tiburón. See Isla Tiburón
Tiger Rattlesnake. See Crotalus tigris
Tohono O'odham. See O'odham Indians (Papago-Pima)
Tohono O'odham Community Action, 204

Toledo, Victor, 5
Torres, Ana Luisa, iv
Torres, Angelita, iv, *160*, 160, 253
Torres, Daniel Armando, iv, *147, 148*
Torres, Dolores, iv
Torres, Isabel, iv, 201–2
Torres, José Ramón, iv
Torres, Manuelita, iv
Tortoises. *See Gopherus agassizii*
Tortuga del monte. *See Gopherus agassizii*
Tortuga de los charcos. *See Kinosternon flavescens; K. sonoriense*
Totoaba, 67, 87–88
Tourism: benefits of, 174, 219; and Seri arts and crafts, 7, *145*, 145–51, 189; as threat to traditional cultures, 28–29
Tozípla. *See Uta stansburiana*
Traditional knowledge, 3; of animals and habitats, 55–59, 66–69, 70, *160*, 197–98; biological value of, 28; of distant places, 49–52; encoded in indigenous languages, 22–29, 38, 204–5, 221; folk tax-onomies, 101–17, 239; hunting, 69, 180–81, 191, 209–10, 231, 235–36, 240, 245; imparting to young people, 216–18; of inter-species relationships, 158–63, *166*, 166–77, 171–73, 228; loss of, 29, 202, 204–15, 206–7; and popular culture, 28; as rationale for indigenous resource management, 183; western science and, 3, 137–38, 170, 173–74, 216, 221. *See also* Language; Medical treatments; Rituals; Songs
Traditional practices, defining, 185
Treaty of Guadalupe-Hidalgo (1848), 139
Tree Lizard. *See Urosaurus ornatus*
Trimorphodon biscutatus (Lyresnake), *65*, *74*, *104*, *169*, *172*, **274–75**, 289; sharing shelters with other reptiles, *66*

Turner Island. *See Dátil*
Turtles. *See* Cheloniidae; *Dermochelys coriacea; Kinosternon*

Urosaurus (brush lizards), *104*
Urosaurus graciosus (Long-tailed Brush Lizard), *73*, *168*, *171*, 286
Urosaurus nigricaudus (Black-tailed Brush Lizard), *75*
Urosaurus ornatus (Tree Lizard), *73*, 96, 112, *128*, *168*, *171*, **262**, 286
Uta lowei (Dead Side-blotched Lizard), 24
Uta nolascensis (Isla San Pedro Nolasco Side-blotched Lizard), 24, *75*
Uta palmeri (Isla San Pedro Mártir Side-blotched Lizard), 24
Uta squamata (Isla Santa Catalina Side-blotched Lizard), 24
Uta stansburiana (Side-blotched Lizard), *73*, *75*, *104*, *128*, *168*, *172*, **262**, 286; carvings of, *149*; Comcáac classifi-cation of, 112; as pets (Efraín), 175–76
Uta tumidarostra (Swollen-nosed Side-blotched Lizard), 24
Utility, reptiles associated with, 125–26, *128*–29
Uto-Aztecan languages, 23
Uto-Aztecan peoples, *51*, 234

Van Devender, Tom, 173, 268, 272
Vaughn, Mercedes, 229
Víbora de cascabel. *See Crotalus angelensis; C. atrox; C. molossus; C. scutulatus; C. tigris*
Víbora de la noche. *See Phyllorhynchus decurtatus*
Víbora de las islas. *See Crotalus estebanensis*
Víbora sorda. *See Pituophis melanoleucus; Trimorphodon biscutatus*
Viperidae (vipers and rattlesnakes), 141, **277–80**, plate 25; carvings of, 108, *149*, *159*; Comcáac classification of, 107, 109; as food, 278; killing

of, 209; material uses of, *145*, 279; medicinal uses of, *138*, 138–39, 278–79; as motif in basketry, *127*; nesting and congregating places, 65–66, 278; ornaments from, 279; sharing shelters with other reptiles, *66*; songs about, 279. *See also Crotalus*
Vipers. *See* Viperidae
Vision quests, *55*, 238, 243, 244

Water holes, protection of, 56–59
Western Banded Gecko. *See Coleonyx variegatus*
Western Coralsnake. *See Micruroides euryxanthus*
Western Diamondback Rattlesnake. *See Crotalus atrox*
Western Patch-nosed Snake. *See Salvadora hexalepis*
Western science: biophilia and, 174; and Comcáac traditional knowledge, 84, 158–59, 167, 170, 173–74, 216, 221; expanded definition of, 158; records of reptile habitats compared to Comcáac, 168–70
Western Shovel-nosed Snake. *See Chionactis occipitalis*
Western Whiptail Lizard. *See Cnemi-dophorus tigris*
Whiptail lizards. *See* Cnemidophorus; Teiidae
Wilson, E. O., 81–82, 122, 152
Women: and butchering of sea turtles, *126*, 186, 236; and taboos against hunting while pregnant, 230

Xantusia (night lizards), *104*
Xantusia vigilis (Desert Night Lizard), *74*, 120, *129*, *169*, *172*, **265**
Xantusiidae (night lizards), **265**; beliefs about, 121; Comcáac aversion to, 121
Xasáacoj. *See Boa constrictor*
Xavier, Juan, 176

Xepe ano cocázni. See *Pelamis platurus*

Xepe ano heepni. See *Crocodylus acutus*

Xepe ano paaza. See *Crocodylus acutus*

Xepe Coosot. *See* Canal del Infiernillo

Xepe Heeque. *See* Canal del Infiernillo

Xica cmotómanoj. See *Dermochelys coriacea*

Xiix haas ano cocázni. See *Trimorphodon biscutatus*

Xpeyo. See *Caretta caretta*

Xtamáaija. See *Kinosternon flavescens; K. sonoriense*

Xtamóosni. See *Gopherus agassizii*

Yaqui Indians, 22; Cahitan language, 50; contact with Comcáac, 50; influence of Catholicism on, 51

Yax quiip. See *Uta stansburiana*

Yellow-bellied Seasnake. See *Pelamis platurus*

Yoemem Indians (Mayo), 22, 23, 234; contact with Comcáac, 36

Yori (non-Comcáac) hunting of sea turtles, 180

Youth: loss of traditional knowledge among, 204–16

Yuman languages, 3

Zebra-tailed Lizard. See *Callisaurus draconoides*

Ziix catotim. See *Gopherus agassizii*

Ziix haas ano cocázni. See *Charina trivirgata; Trimorphodon biscutatus*

Ziix hast iizx ano coom. See *Sauromalus obesus*

Ziix hehet cöquiij. See *Gopherus agassizii*

Ziix tocázni heme imocómjc. See *Dipsosaurus dorsalis*

Ziix xepe ano quiih. See *Chelonia mydas*

Zostera marina (Eelgrass), 65, 167, 179, 218, 240, 244, 245

Designer:	Kristina Kachele
Cartographer:	Bill Nelson
Compositor:	Integrated Composition Systems
Text:	10.4/15.5 Monotype Joanna
Display:	FF Eureka Sans, Journal
Printer and binder:	Malloy Lithographing, Inc.